Troubled Natures

Waste, Environment, Japan

Peter Wynn Kirby

University of Hawai'i Press
Honolulu

Library of Congress Cataloging-in-Publication Data
Kirby, Peter Wynn.
Troubled natures : waste, environment, Japan / Peter Wynn Kirby.
p. cm.
Includes bibliographical references and index.
ISBN 978–0–8248–3428–9 (hardcover : alk. paper)
1. Human ecology—Japan. 2. Japan—Environmental conditions—Public
opinion. 3. Waste products—Japan—Public opinion. 4. Pollution—Japan—
Public opinion. 5. Public opinion—Japan. 6. Urban ecology (Sociology)—
Japan—Tokyo. I. Title.
GF666.K57 2011
304.20952—dc22
2010026342

Designed by Wanda China
Printed by Integrated Book Technology, Inc.

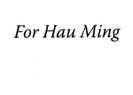

For Hau Ming

Contents

Acknowledgments

During three years of ethnographic fieldwork on waste, pollution, vermin, "outcasts," and other unsavory-seeming topics, I was sensitive to the fact that many Japanese found this to be a rather odd way for a member of their elite Tokyo University to be conducting himself. One gentleman who lived nearby kept trying to introduce me to practitioners of the traditional arts and other refined pursuits, reasoning that if my discipline was "cultural anthropology," he would steer me (away from the gutter, I suppose) toward Japan's more edifying "high" culture. While amused friends, neighbors, and acquaintances eventually resigned themselves to my strange-seeming research, I hope that other Japanese, when they read this book, will see it as a determined effort to understand thorny environmental questions in contemporary Japan from a sustained sociocultural perspective, and not as an American's attempt to air Japan's dirty laundry. Indeed, perhaps this work will lay the foundation for a wide-ranging analysis of America's notorious ambivalence on environmental issues.

This research has been a long time in the making, and I would first like to thank my two supervisors at Cambridge, Keith Hart (for the M.Phil.) and James Laidlaw (for the Ph.D.). Keith read several draft chapters of this book, offering trenchant commentary, and has been extremely supportive, blunt, humorous, and helpful generally, particularly while I was based in Paris recently. James, for his part, guided me through the doctoral labyrinth and was particularly good at interrogating some of my reflex environmentalist attitudes and assumptions about Japan, which greatly helped position the research project and the final Ph.D. thesis. Roger Goodman was my external examiner, from Oxford, and helped steer the book project. Alan Macfarlane was generous with his time during my years at Cambridge. Jerry Eades was an early and enthusiastic mentor, promoting my research and helping me in numerous ways. Mitch Sedgwick, now in the office just next door here in Oxford, has long been an eager participant in wide-ranging discussions on Japan and the peculiar joys and anguish of anthropology and academic life generally. Patrick Beillevaire, director of the Centre de Recherches sur le Japon at the EHESS in Paris, gave me the institutional space to expand on some of this waste-related research. My brother Rob Kirby gave several key chapters an exceedingly careful and considered reading. My dad, Bob Kirby, trained his prodigious ornithological expertise on the chapter on verminous "urban jungle" crows. Anne Allison and Marilyn Strathern also read chapters and gave advice at crucial moments in the writing of this monograph. Fascinating honors seminars, taught by James White and Jane Bachnik, respectively, reignited my interest in Japan while in college.

In Japan, Funabiki Takeo was my supervisor at Tokyo University, selflessly opening doors and making introductions (notably to top editors at TV-Asahi after a journalistic scandal there), and very thoughtfully facilitated affiliation in the department for later postdoctoral research in Tokyo. Yamashita Shinji, another colleague at Tōdai, was also helpful in providing early fieldwork guidance. Lieba Faier, Miyoshi Mika—two of my anthro pals at Tōdai—and I helped each other get through the peaks and troughs of ethnographic fieldwork and grad-student existence. Steve Jarvis and Mike Faul (members of my Tōdai Monbushō cohort, along with Chris Spang) offered their hospitality during field trips and ready companionship during adventures throughout Japanese society. Azuma Kenichi obligingly provided information on dioxins and incinerators in Japan at short notice. Kumada Naoko, a Cambridge classmate, was a source of balanced counsel on things Japanese. The expert staff at the fabled Inter-University Center in Yokohama, especially the exacting Sano-sensei, sharpened my language skills considerably. David Askew, Rie Askew, Ishī Yuka, Edmund Thompson, and Yu Mi-Jeoung helped me survive the rigors of working in a Japanese university and the demands of lecturing in Japanese to hundreds of students each week. These acknowledgments leave out hundreds of Japanese people who patiently met with me, discussed environmental problems and contemporary Japanese society, sat for formal interviews in their homes or neighborhoods (or sometimes in the back of a distant café, depending on the need for privacy), and/or allowed me to participate in community activities, leisurely pursuits, domestic tourism, activist initiatives, environmental hearings, and so on. Some of these Japanese would welcome going on the record, but in order to ensure that those who wished to remain totally anonymous could do so, I have taken extensive precautions to make sure that places and identities remain obscured. This included scrubbing geographical references and institutional phone numbers out of some waste-related signage that appears in this book. (All photographs are my own.) Japan can be a warm, captivating, even magical place, and I thank all the people I know there for their acceptance, generosity, and good humor.

My research received the generous support of an Allen Scholarship (University of Cambridge), a Monbushō Japanese Ministry of Education scholarship (held from 1998–2000 at Tōdai), a grant from the William Wyse Fund (administered by the Council of Trinity College, Cambridge), and a grant from the Audrey Richards Fund (Department of Social Anthropology, Cambridge). While working as a tenured assistant professor at Ritsumeikan in Japan, I also received munificent financial support from the Ritsumeikan Trust, along with ample freedom to travel and conduct further field research and library study in Tokyo and abroad.

My thanks go to the professional University of Hawai'i Press team, in particular the talented Margaret Black, who passed away toward the end of the process. Margaret's keen eye, her ear for language, and her unlikely interest in waste issues improved the monograph considerably.

Finally, Hau Ming Tse has slaved over cover art and book illustrations late into the night, endured periodic absences and the general existential weirdness of anthropologists, parachuted into Japanese communities at times of great stress with almost no language skills, and tolerated an unruly pile of books and papers on the dining room table, to name but a few challenges and sacrifices. This book is dedicated to her.

Note

This book uses macrons to indicate long vowels, but it leaves words that are already common in English in their more familiar form. Therefore, Nīgata (the city on the Sea of Japan, pronounced "KNEE-gata") is written with a macron, while Japan's capital is still written as "Tokyo." Throughout the book, I follow the Japanese convention of writing Japanese names with the family name first.

FIGURE 1   Detail of a hand-painted sign emblazoned with the word "Garbage!" *(gomi)*, a warning against illegal dumping outside of Tokorozawa.

CHAPTER 1

Introduction
Japan's Waste Shadow

Our world is engulfed by discourse on environment. While a discrete realm of study in its own right, the environment has also become the backdrop to what many people think about when they conceive of the present, the future, and the recent past. Yet amid all this environmental attention, the variety and complexity of conceptions of "environment" remain frequently downplayed, even ignored. Throughout the world, how surroundings are conceived, invoked, and enacted is highly contextual. Nevertheless, still pervasive is the notion that environmental conditions, perhaps like meteorological conditions, are more or less universal and therefore comparable—one hotspot, rainforest, city, or nation can be measured and compared with another.

Troubled Natures confronts this problem, interrogating what "environment" really *means* in one of the world's most important and complex non-Western societies: contemporary Japan. Postwar Japan, onetime conjurer of the economic miracle, has been widely reviled for creating environmental debacle. This reputation has been well deserved, by and large; yet environmental problems in Japan are, significantly, embroiled in exceedingly Japanese ways of conceiving, relating, and interacting. Knowledge of this can become a powerful tool. This volume probes into environment in Japan—primarily by focusing on the foul stratum of waste—in order to convey the breadth and subtlety of these questions, which in turn have important ramifications for how environmental threats are tackled in a wide range of sociocultural settings. Waste and pollution are, in many ways, particularly well suited to this wider discussion: at first glance they obviously comprise a key environmental issue, but upon greater scrutiny they offer a revealing cross-section of varied social elements in contemporary Japan. That

Japan—highly urbanized, industrialized, and consumption-oriented—is in many ways on the front lines of contemporary waste policy makes this study particularly apt.

Environmental issues are rarely separate from wider social concerns, as the case of Japan demonstrates. I embarked upon long-term fieldwork in Tokyo in 1998 and worked extensively in Japan through 2006, conducting further field research in 2007 and 2009. When I first arrived, the Japan I found was struggling and dispirited. Long accustomed to high-speed growth, the nation was bogged down in recession. Workers used to lifetime employment were being forced into retirement. Bankruptcies, once all but unheard of, became pandemic almost overnight. High-profile financial and political scandals plagued the nightly news. Pillars of Japanese social stability—cultural "homogeneity," low crime, shared sacrifice, respect for hierarchy and tradition—were becoming destabilized, with younger generations questioning and at times undermining the very foundations of Japanese solidarity. The social repercussions of this upheaval were seismic, and a number of Japanese I encountered were unusually frank about this difficult period, whose effects endure even today. Informants constantly transposed "economic," "social," "political," "family," and "environmental" commentary in articulating their opinions and beliefs regarding waste and illness—and life generally. This book therefore attempts to sketch, at varying scales, the range of sociocultural issues that encompass "environment" in Tokyo and beyond in order to chart Japan's wastescape with the nuance it requires.

This ethnographic journey led in unexpected directions and into controversial territory. Initially a study of the contradictions inherent in urbanized, industrialized, even polluted Japan's fixation on "nature" (with its bonsai trees and manicured gardens), my research uncovered new challenges besetting Japanese society. Most had roots in Japan's toxic wastescape: waste-borne reproductive threats and sexual anxieties in a society with a stubbornly declining birth rate; widespread concerns over environmental illness caused by waste facilities in particular and pollution generally; state attempts to cultivate Japanese "sustainability"; and tensions between state prerogatives and local welfare in handling contaminated communities. The project also led deeper, to buried social and cultural underpinnings of Japanese society—for example, racist self-imaginings and wartime practice, along with contemporary outcasts, social exclusion, and ostracism—all of which are bound up in ideas of waste and pollution. This volume, thus, interprets the social and environmental challenges of a complex, non-Western society via a penetrating cultural analysis of waste that unearths unexpectedly troubled natures at the heart of Japanese life.

The Shadow Archipelago

Four brief sketches give a sense of the ubiquity and topicality of waste in contemporary Japanese society:

As befits the "Land of the Rising Sun," the acme of Mt. Fuji in summer is packed with hundreds of hikers huddled against the bitter cold and high winds at 12,388 feet (3,776 meters) to view the sunrise, the first moment the sun touches Japan. Though Mt. Fuji is regarded as a sacred peak, and holds unparalleled significance in Japanese culture, the surface of the mountaintop is encrusted with drink cans, energy tonic ampoules, empty water bottles, wrappers, and other detritus dropped by exhausted climbers who squeeze between the improbable noodle shops, vending machines, and souvenir stalls that crowd the summit. As the dense stream of bodies, four or so abreast, shoulder-to-shoulder, trudges up the main trail, more climbers are able to reach the peak's volcanic summit and consume its amenities while returning hikers spill down the various descents like human lava.

* * * * *

During a routine safety check at a Tokyo community pool, three Japanese swimmers, two women and one man in their late twenties, sit on the pool edge and confess their discomfort with the fact that the pool water is heated by recycled energy from a major waste-incinerator complex—in the middle of which the pool lies—whose enormous smokestack looms hundreds of feet above the swimming area. The community has been bombarded for months by media reports regarding the dangers of toxic pollution from incinerators in Japan.

* * * * *

At a marina along the coastline, tossed cans, plastic bottles, and other flotsam bob in the crevices of massive concrete tetrapods that create a breakwater there. On the long path to the parking lot, a middle-aged Japanese man, finding a pile of waste abandoned by yachters and others—public trash cans are extremely scarce in

Japan—pulls out a lighter and starts burning the heap of plastic bags, food wrappers, Styrofoam trays, and leftover food garbage to protect against vermin, thereby creating a plume of acrid smoke that soon envelops the jetty.

*   *   *   *   *

Four men tee off at a golf course on an artificial island in Tokyo Bay. Despite their good-natured grumblings, the men are forbidden to smoke on the course due to the highly flammable gases wafting up through vents from the millions of tons of waste decomposing beneath the manicured fairways. In the distance on all sides, the expansive bayscape surrounding them is filled with similar islands made of garbage—an "inland sea" of wastelands that provide valuable real estate for infrastructure and development.

Japanese postwar society, driven by mass consumption and massive construction, is enmired in waste. Flows of waste that circulate throughout mega-cities such as Tokyo warp the ecology and pollute the living environment of millions of urbanites. Furthermore, social reckonings of waste and pollution influence environmental conduct and social action in a variety of arenas. The fallout from this frequently toxic waste is every bit as political as it is "environmental," and waste issues expose fault lines in Japan's contemporary life that resonate with community friction. One novel way of understanding contemporary Japan—seen here through the prism of Tokyo—is to delve into what Japanese dump, burn, recycle, or hide: the unsavory, and often polarizing, sphere of waste.

Waste, though ubiquitous, certainly need not be pedestrian. There are numerous ways to measure and analyze waste, and scholars in geography, urban planning, biology, and related disciplines have produced numerous important perspectives, most of them empirically based. Yet properly assessing how humans interpret waste and pollution requires a more comprehensive definition of waste and a more inclusive, ethnographic research focus. Attitudes toward waste are not universal; they modulate with reference to history, differences of language, notions of productivity, constructions of thrift, regulatory landscapes, classification of vermin and outcasts, exposure to and participation in local, regional, and global environmental discourses, and so on. Even within a given society there are marked variations. The common practice of viewing waste as a universal "environmental" problem, and regarding environmental practices in vastly different societies as comparable variables, leaves important elements of waste consciousness out of the equation. In order to account for the depth, ambivalence, and complexity of the subject, *Troubled Natures* engages not only with familiar and unsavory forms of waste produced in cities—such as solid organic and toxic waste and different forms of harmful air pollutants, including dioxins[1]—but also with topics that do not normally fall within the purview of environmental accounts. These include the ways in which ideas of waste become entangled in socioculturally maintained boundaries of hygiene, health, and illness; status-laden and often exclusionist constructions of purity, identity, and nationhood; and the profound moral dimensions of waste and excess in a time of mass consumption and mass disposal. Deeper and more diffuse than the tensions between state bureaucratic fiat and local welfare that perplex many societies, waste issues in Japan unearth profoundly serious broader controversies. These concern perceived "social pollution" and untouchability at a time when many Japanese publicly eschew visible ostracism; anxieties surrounding reproduction and fertility in a society with a stubbornly declining birth rate and living environments that can be toxic and hostile toward procreation; and debate over the costs and benefits of pro-growth economic policy amid social fragmentation and environmental damage.

Tokyo serves as an important ethnographic stage. Like all cities, Tokyo is essentially an aggregation of flows. Due to its size, its cultural position as a non-Western world capital, and its mighty (though sometimes recessional) economic status, Tokyo's flow-channels can be even more extensive and influential than one might imagine. For example, as Theodore Bestor (2001) has demonstrated, Tokyo's mammoth Tsukiji Fish Market stands as *the* key node in the worldwide tuna trade, and by extension in the pricing and trading of most premium sushi-grade and other fish internationally. (Should the politics of the International Whaling Commission keep moving in less conservationist directions, perhaps this will be the case for whale sashimi as well in another decade or two [see Chapter 8].) As the capital of Japanese industry and as a magnet for intensive domestic housing and commercial construction, Tokyo also has considerable influence over clear-cut logging of Southeast Asian rain forests, with the complicity of often overseas-Chinese logging concerns (Dauvergne 1997), and impacts the forest ecology of other areas, such as Siberia. Clearly, then, some flows are not symmetrically reciprocal. The fact that the Tokyo-Yokohama-Kawasaki conurbation accounts for nearly a quarter of Japan's population of 127 million (e.g., Cybriwsky 1998) and remains an industrial and postindustrial powerhouse points to an unsustainable imbalance between human development and material resource consumption that has had a destabilizing effect on the Japanese archipelago as well as on regions far beyond the boundaries of Japan. But it is nevertheless clear, given the above, that a snapshot of "Tokyo" obscures numerous fluid and emergent elements of life that contribute much to the capital's character. Along with those movements of capital, labor, and culture more commonly focused upon in analyses of "global" cities (Sassen 1991), labyrinthine circuits of resources, waste, pollutants, and disease create an undercurrent of defilement, contamination, and illness that is nevertheless rich in the fundaments of sociocultural research.

Interpreting the Wastescape

Consider the contours of Japan's wastescape. Contemporary Japan, the world's second-largest economy, with a society largely organized around conspicuous consumption, is a macroscale work in waste production. Tokyo serves as an evocative case. Whether one confines Tokyo to the core twenty-three wards of the capital or includes outlying prefectural zones as well as Yokohama and Kawasaki in one's calculus, the daily waste output of "Tokyo" is massive. To give some idea of the scale of this urban disposal: in 2002 Tokyo produced 5.17 million tons of domestic and commercial waste, of which the core wards produced 3.82 million tons, and industrial waste for Tokyo Prefecture reached 23.5 million tons (Tokyo Bureau of Environment 2006a). Naturally, this waste takes myriad forms (some materializing as compounds whose structural characteristics, let alone toxic effects, have not yet been identified) and is dispersed variously, including through air emission, water contamination, and visible litter. Many thousands of truckloads of domestic, commercial, and industrial waste are piled each year into incinerators, whose plumes of emissions disperse across the city, the prefecture, and even across Japan's national borders. These infiltrate Tokyo's human and ecological environments with great tenacity. Approximately 80 percent of Japan's waste is incinerated (Government of Japan 2006), a space-saving disposal strategy that the state eagerly pursues. What is not burned is dumped by yet other trucks to create artificial landfill islands or land-reclamation projects that have transformed Tokyo Bay and extensive stretches of Japan's coastline. But often as well, waste loads are carted off to distant rural communities willing to accept piles of urban waste to fill their municipal coffers. Increases in waste regulation and the levying of fees means that illegal dumping occurs frequently, particularly on the margins where perpetrators can escape scrutiny. What verdant fastnesses remain near conurbations easily fall victim to such activity, offering conspicuous proof of the wastescape's reach in precincts far from the capital (figure 2). Other dumping is more ambiguous: witness urban owners' abandonment of pets such as cats and dogs on the periphery, leading to an abrupt rustication of these domesticated pets in a simultaneous dilution of "the wild" (figure 3). Impromptu incineration of wasteheaps in desolate areas, urban alleyways, vacant lots, and backyards, now illegal but still not so infrequent, contributes to the air pollution wafting around human agglomerations. The scale of Tokyo's wastescape—or as Peter Dauvergne (1997) might say, the ecological "shadow"[2] that Tokyo casts over the rest of the nation, over East and Southeast Asia, and over much of the Pacific Rim—is therefore vast enough to place waste issues prominently in the topography of contemporary Japanese society.

Any conceptualization of Tokyo's *waste shadow* must be broad, accounting for a complex of waste flows and waste avatars across a varied sociocultural terrain. Though the waste shadow I describe bleeds across the boundaries of "Tokyo" and raises cartographic issues, this analytical dimension differs categorically from the superimposition of arbitrary layers over an objective map, as with color-coded "political" and "topographical" cartographic representations familiar to map readers. Rather, this wastescape is a subjective field, shaped more by sociocultural attitudes than by territorial features such as rivers, coastline, mountain ranges, and administrative boundaries. Certain hotspots in a region may loom immensely in the imagination of inhabitants due to the plumes of toxicity they belch out—or are perceived as belching out—as I show with a contested waste facility located in one of my Tokyo fieldsites, just as other areas may diminish in significance, whether justified by environmental technoscience or not. This terrain, furthermore, is very much shaped by media depictions. One alarmist broadcast on toxic pollution can, overnight, reverberate throughout an entire wastescape, as happened in Japan in 1999, when a television news report transfixed the nation and buffeted consumption patterns for months. Or, with a relative decline in anxiety over a certain form of waste, such as automobile exhaust, the waste shadow may gradually transform to reflect this change in perceptions. The wastescape I negotiated over a number of years in Tokyo went through several such dramatic metamorphoses as well as other shifts that were less discernible to a casual observer.

While many issues here bring into question the wisdom of think-

FIGURE 2   Detail of a sign warning against illegal dumping outside Tokyo.

公園に、子猫を
捨てないで下さい。

FIGURE 3   A park sign, with graffiti, warning Japanese not to "throw away" *(suteru)* kittens there, using the same verb as is used for throwing away garbage.

ing of the world entirely in terms of politico-administrative/territorial units such as nations or metropolitan districts (see Kirby 2009a), operations of production, consumption, and waste make a particularly convincing case for scholarly reconception of arbitrary boundaries between cities and their surroundings. The sheer volume of human and material resources drawn or transported to the Tokyo area, along with the millions of tons of solid, liquid, or ambient waste emanating from there to outlying or far-distant zones, blurs distinctions between the capital and its wider environs so much that their separation can be misleading, even intellectually dishonest. Waste in all its material forms comprises an often willfully ignored, but unmistakable, overlap between the "human" and the "environmental," which makes overly rigid segregation of societies from the "nature" they conceive around them difficult to sustain. The discards, residue, and exhaust that accrete to the urban periphery make distinguishing the one from the other all but impossible. Yet there is, admittedly, widespread significance to Japanese of conceiving of localist group membership very much in terms of such administrative districts. In the case of waste controversies, divisions between "us" and "them" lie at the core of environmental disputes. The makeshift armada of private garbage trucks that transport waste from the capital and dump it in seeping landfill in surrounding communities—accompanied by tensions between local prefectural management and centralized bureaucratic dictates—lead not infrequently to suspicion or resentment of the capital in the many communities affected by Tokyo's waste disposal policy. Analysis of waste in Tokyo must therefore remain attuned to this issue of scale, with the miasma of waste spilling into the surrounding ecology creating a backdrop for understanding how Japanese urbanites and their rural neighbors grapple with waste issues on a variety of sociocultural levels.

In these pages, *Troubled Natures* excavates the sometimes deeply embedded notions of waste that extend throughout Japanese social life from conspicuous consumption and thrift to outcasts and vermin. But the political valences of waste in Japan, and the complex social resonances of environmental issues, beg a thorough analysis of ideas of nature in Japan and their influence on identity, anxiety, and action there.

Japanese "Natures"

Japan looms large in a world perplexed by environmental questions. Its erstwhile reputation as Environmental Enemy Number One has largely been usurped by America and China since the turn of the millennium. And yet Japan's role as a perennial villain (so cast largely by international mass media and global environmentalist groups, and with good reason) hovers uneasily near the idea of Japan as seductive, refined, and exotic. Amid images of depleted Southeast Asian rainforests, slaughtered whales, and exported or outsourced pollution, Japanese society's embroilment in transnational environmental debacles of various registers commingles with tableaux of paper houses, manicured gardens, miniature bonsai trees, and other ready images of natural-cultural engagement to create a disjunctive impression of Japanese environmental behavior. This ambivalence is mirrored in the self-imaginings of many Japanese as well, though most discern contradictions in their thinking only after long discussions on the topic.

Indeed, one striking feature of contemporary life in Japan remains the contrast between Japan's image of itself as a society that embraces nature and, conversely, contemporary examples of pollution, hyperurbanization, and environmental intervention there. While many Japanese believe nature and its experience to be at the core of their Japaneseness, they also, generally speaking, widely tolerate and at times participate in defilement of the very natural milieu they prize. That Japanese focus so much on the poetry of impermanence in natural settings has become, to an extent, a double-bind; for this fixation on fleetingness, on the surface a sign of deep, society-wide sensitivity and respect, contributes to a form of detachment that pervades the Japanese context. As I illustrate, the very transience of nature that Japanese frequently treasure (the ephemeral moment of perfection in a pied scene of autumn foliage, the brief period when cherry blossoms reach full bloom) contributes, to a degree, to a relative lack of concern with environmental destruction. Indeed, the more such sublime examples of nature-in-action can be considered to be "vanishing" (Ivy 1995), the more they become prized and then aggressively consumed. It must be said, however, that since the early nineties in particular, many Japanese have begun to tap into elements of the global environmentalist discourse to the extent that in recent years they have begun to view pollution and hyperdevelopment in clearly negative terms.[3]

Japanese mass media play an important role in cultivating and reinforcing these and other notions. A deluge of messages and advertising—engulfing metro and rail stations, roadside billboards, mail slots, mobile phone displays, television and radio broadcasts, Internet browser windows, disposable wrappings, the vertical face of stairs, escalator handrails, and plasma-screen monitors installed in subway and train carriages, coaches, and even automobiles—immerses Japanese urbanites in a "sea of signs" (Yatsuka 1990) that subtly influences and commodifies constructs of the natural. Arresting natural scenes are sometimes used to communicate social harmony, optimism, and supposed purity of motive—on the part of banks and corporations, for example—as well as a gritty determination to root out corrupting influences (see figure 4). Yet this same media apparatus transmits images of pollution and environmental degradation and actively fans anxieties over community toxicity, environmental illness, contaminated breast milk, low sperm count, infertility, birth defects, climate change, foreign criminal elements, the perceived lost moral compass of Japanese youth, and a host of other powerful issues. All this makes the responses of Japanese urbanites with regard to "nature" exceedingly complex and politically valenced. One important dimension of my research was tracking microprocesses of local discourse as informants received, interpreted, and retransmitted mediated knowledge, frequently embellishing it with their own opinions and insights. Though "global" attitudes toward environmental issues

FIGURE 4   A public service message, posted in Tokyo's Shinjuku Station, warning: "We don't need organized crime *(bōryokudan)* in our community." The panoramic natural backdrop to this message depicts a lake in a noted Hokkaidō tourist destination, the Shiretoko lake district.

seeped into Japan throughout the years of my field research, the degree to which members of communities adopted the terms of the global environmentalist discourse depended in a number of cases on the degree of their exposure to risk and their trust in local, metropolitan, and national governmental competence, the latter leaching away in a decades-long flood of corruption scandals and well-publicized cases of mismanagement of local environmental policy there.

Clearly, then, ideas of nature are entangled not only in the environmental thinking of Japanese urbanites but in their views on the shifting sociocultural topography of contemporary Japan.[4] The polysemic nuances of the word "nature" in English convey rather neatly the interrelationship between being and surroundings in human social worlds [OED 1989; Williams 1988].[5] On the one hand, a person's "nature" derives from the complex of sociocultural forces that shapes identity and behavior and worldview. On the other, "nature" betrays a cultural conception of surroundings, often (though not always) distinct from human settlement, that operates as a mirror of a society itself—that is, nature is not an ecological given, as is frequently assumed, but instead reflects the cultural-historical dispositions of a particular society and differs in important respects from ideas of nature constructed in another social milieu (Berque 1997a; Asquith and Kalland 1997).[6] That there is an important correspondence between the "natures" of members of a society and the "nature" they perceive around them is far more than a play on words. Indeed, the relationship between being and conceived surroundings is fundamental. In contemporary Japanese rhetoric, the links between "natures" and "nature" can take on an additional, darker, shade of political meaning as conservative thinkers invoke the "climate" of Japan and certain sociohistorical traits as the central ground of Japanese uniqueness vis-à-vis less cultivated, even "barbarian," outsiders. In this racist, nativist construction, the correspondence between nature(s), natives, and nation, all etymologically and politically entwined, could not be more direct. While we do not have to accept these ideologues' crass, just-so reasoning, the wide diffusion of their arguments throughout Japanese society (Dale 1986; Yoshino 1992; Befu 1993, 2001) influences the social logic of contemporary Japanese to such an extent that this base register of discourse cannot be ignored.

This book plumbs deeply into the "troubled natures" that I found in my Tokyo fieldsites and in discourse in Japan more generally. My core fieldwork compared environmental consciousness and environmental conduct in two communities in Azuma Ward,[7] Western Tokyo. The first, Izawa, was riven by a toxic-pollution protest against a waste facility located in the center of the community; the second, Horiuchi, was a sleepy, rather ordinary-seeming district that—like many in Tokyo—found itself bombarded by mass media reports regarding toxic pollution, reproductive threats, environmental illness, and so on during the time I lived there. In each community the tension between lofty rhetoric and daily practice helped highlight the practical ambivalence of Japanese environmental consciousness, whether the rhetoric stemmed from idealized views of Japanese harmony with nature chafing against the realities of everyday life in the wastescape or whether the rhetoric was inspired by global environmentalism, positioning the ideal of sustainable, salubrious environments against the economic imperatives of post-Bubble Japan.

I then examine broader social issues—a vermin epidemic, exclusionary practices, recycling efforts and lifestyle changes, toxic threats, and questions of identity—and trace links between my fieldsites and larger processes elsewhere. Throughout, I found it impossible to examine waste and environmental health problems in Tokyo without colliding against all manner of diverse cultural elements linked to nature(s)—uneasy relations between animals and humans, "native" conceptions of the foreign and the polluted, selective and labile environmental priorities, reproductive challenges in the face of a plunging fertility rate, and changing attitudes to illness and health. Inquiry into the ways in which environmental questions circulate throughout Japanese society furnishes insight into central elements of contemporary Japanese life.

As the title to this book suggests, these interwoven sociocultural strands—all entangled in some way with "nature"—are variously vexing, contested, and under negotiation. But while attentive to the undeniable significance of conceptions of nature, this volume is also "after nature" (Escobar 1999; Strathern 1992). It analyzes a self-proclaimed nature-focused society, Japan, with an eye toward key contemporary social developments (waste, environment, energy, sustainability) without being overly fettered by the analytically limiting and socioculturally anchored concept of "nature."

Relating Waste

Waste envelops human endeavor. Every process of growth brings decay. Every act of production yields unwanted byproducts. Every act of consumption produces residues. While these "laws" of waste can be considered universal, it is how societies deal with waste that makes waste an important focus of sociocultural scrutiny. Waste, like dirt, is often considered to be "matter out of place" (Douglas 1966, 35) when not segregated in containers or zones (bins, toilets, dumps, composts, prisons, ghettoes, and so on) designed to sequester it from daily view. In short, waste is that which societies typically wish to hide. If only for this reason, waste becomes an important stratum of sociocultural artifact, perhaps overly redolent of social abandonment and decay but illuminating with regard to discerning what social groups seek to obscure.

Waste generally falls into three overlapping, interpenetrating categories of social and semantic resonance. To start with, waste is residue. Whether intentionally or not, and for all its ability to change form, waste stubbornly remains. At times, this byproduct is neutral or even desirable—take grappa, for instance, made from the leftover pomace (the skin, stems, and seeds) of pressed Italian wine grapes; or consider the oxygen released through photosynthesis as plant life consumes carbon dioxide from the earth's atmosphere. But more often waste gets in the way despite the best efforts of social actors to rein it in or ignore its presence entirely. Though waste need not be a menace, clearly it undergoes a conceptual transformation when it reaches a certain level of appeal or utility, at which point it may no longer be viewed as common "waste" at all. Sometimes it is the perceived benignness of one form of waste compared to reviled alternatives—say, fossil fuel or nuclear power—that helps imbue the waste-product with an aura of desirability. The natural gas methane is a case in point: present in the earth's crust and in the belching and flatulence of cattle, it is, in its varied forms, a relatively underexploited fuel source.[8] Though waste can comprise a clear, relatively neutral conceptual category as residue, its designation in social usage frequently implies an assessment of value. Such reckonings of value can change, particularly when aided by clever design, repackaging, and marketing skill—take, for example, Patagonia Synchilla fleece clothing made from recycled PET plastic found in discarded mineral water bottles, among other sources.

Next, and perhaps most visibly, waste is litter. Clearly, the divisions between residue and litter are permeable. But location, as they say, is everything. While some may tolerate the profusion of certain forms of waste in contemporary social life, waste perplexes when jettisoned in unacceptable quantities or in inappropriate places. Such judgments of "acceptability" and "propriety" are, of course, socioculturally anchored and not often universally shared even within a given sociocultural context. In many cases waste, like dirt, "exists in the eye of the beholder"

(Douglas 1966, 2). An area strewn with fallen branches after a storm may seem unkempt to some but abundant in easy firewood for others. Waste "out of place" is an assault on social order, and the degree to which communities and states mobilize to combat this threat furnishes important ethnographic data on the relationship between waste and control.

In a third, linked register of nuance, waste is excess. In every society there is a threshold past which ordinary ingestion becomes surfeit, when consumption becomes overindulgence, where action becomes overkill. In English, this connotation is neatly encapsulated in the same term, "waste," that designates the remainder: some consumption is construed as a "waste," a profligate extravagance. But beyond this easy lexicographic coherence, there are numerous senses in which material waste (garbage, trash, refuse, rubbish, offal, detritus, ruin, what have you) and related phenomena are embroiled in the moral judgments that characterize discussions of improper excess from an environmentalist perspective. Succinctly put, environmental waste has become an issue of character and, by extension, of reputation, as individuals, communities, corporations, and governments seek to demonstrate care for ecological surroundings in a world ringing with global environmental discourse (or not, as the case may be, as shown by largely contrarian regimes in China, Russia, and even the Bush-era United States). As this volume demonstrates repeatedly, the notion of waste is hardly limited to a putative "environmental" dimension. Drains on public resources (and public patience) of one form or another—such as, depending on one's perspective, the "feckless" unemployed, "irresponsible" welfare mothers, school dropouts, "benefit-hogging" asylum-seekers, regressive tax cuts, young women who refuse to marry and procreate early—as well as the numerous cases of conspicuous (or aberrant) consumption that go beyond the pale are often phrased in terms of the waste metaphor. The worthless, the spurious, the filthy, and the flawed are regularly condemned with terms such as "garbage," "trash," and "rubbish." Baudrillard (1998) critiques the worship of affluence and prodigality that societies based on scarcity exhibit. He, indeed, poses the question of whether *all* production is waste.[9] Clearly, then, reckonings of waste can differ markedly and their decryption constitutes an important analytical task.

Describing Waste

Waste is a complex topic, and there exists some unhelpful terminological ambiguity in its scholarly analysis. Several writers have chosen terms they believe reflect the essence of waste and its many corollary notions. Yet semantic tensions arise due to the divergent scholarly aims of the authors. For example, John Scanlan (2005) exalts "garbage," while Michael Thompson (1979)—later joined by William Rathje and Cullen Murphy (2001)—prefers to speak of "rubbish."[10] Scanlan's terminological decision is driven primarily by his philosophical focus on that which is actively segregated and edited out. He explains in his insightful book *On Garbage* that "the creation of garbage is the result of a separation—of the desirable from the unwanted; the valuable from the worthless, and indeed, the worthy or cultured from the cheap or meaningless" (Scanlan 2005, 15). Yet the term "garbage," like "trash" and "rubbish," oozes with the condemnation that critics heap on "vile" pornography, "bad" writing, "worthless" art, or "spurious" theories, and this pointed social usage, I believe, undermines its utility, as I explain below.[11] (Scanlan claims that "garbage" is the more general term, with "waste" a more particular variant, but to my ear the reverse is true.) For landfill archaeologists Rathje and Murphy (2001, 9), the decision was influenced by precise definitions that have fallen out of correct usage:

> *Trash* refers specifically to discards that are at least theoretically "dry"—newspapers, boxes, cans, and so on. *Garbage* refers technically to "wet" discards—food remains, yard waste, and offal. *Refuse* is an inclusive term for both the wet discards and the dry. *Rubbish* is even more inclusive: it refers to all refuse plus construction and demolition debris.

They offer no rationale for rubbishing the term "waste," but presumably their decision was influenced by the exclusively material nature of the waste on which they focus and by the discourse of landfill management. Rathje and Murphy's archaeological digs in urban landfill regularly unearthed putrid, even liquefied, "finds" from past decades, so their terminological precision bears logical utility in avoiding confusion. Martin Melosi's authoritative *Garbage in the Cities* (2005) also adheres to strict definitions,[12] though he periodically finds reason to diverge from these.[13] Significantly, Melosi (a long-standing authority on the subject in the discipline of history) nevertheless reverts to the term "waste" time and again, both in facilitating specific definitions and in generalized discussion. Kevin Hetherington (2004) prefers to speak of "disposal," which is an important facet of consumption theory but which seems less well suited to a wide-ranging ethnography of the wastescape. And though

Douglas' influential theories (1966) on "dirt" and "pollution," which I take up in Chapter 6, have had a noticeable impact, these variations, terminologically speaking, seem too limiting to be of use here.

This ambiguity and complexity influenced my terminological stance. "Waste" offers the broadest lexicographic scope in addition to its tropic dexterity, and it comes free of many of the other terms' unwieldy connotations. In a sociocultural analysis of contemporary Japan sensitive to environmental dimensions, the clear moral register of waste, scorning excess, actually complements the study somewhat rather than distracting from its aims. Admittedly, "waste" can be dry and scientistic, leaving moral nuances underexamined, but properly defined and handled, I believe the term captures the many important cultural and semantic nuances of exhaust, discard, and excess, not to mention their numerous interlinkages in consumption-oriented capitalist societies like Japan, Europe, or America. And finally, my study is less concerned with distinguishing specific types of material waste, unlike the important (albeit odoriferous) subdiscipline of landfill archaeology, so I freely use "waste" in place of other terms to cohere the varied forms for analytical purposes. (This is a judgment that other social theorists have made, for example, Baudrillard [1998].) In later chapters, I delve into native Japanese terms for waste, for which this English term is the best general translation, and explain their resonance in my fieldsites.

Waste, "Risk," Sustainability

One reason waste inspires contemporary misgivings is that its dangers, real or perceived (Douglas and Wildavsky 1982), can extend considerably further than the trash can or dumpster. Throughout much of history, waste was an inconvenient mundanity and a particularly aromatic omnipresence in urban settlements. From ancient times through to the early modern period, urban denizens of most societies cast waste into the streets and "waste collection," if it existed at all, consisted mostly of scavenging (Strasser 1999).[14] Sometimes, as Melosi (2005) explains, the litter became so voluminous that communities simply rebuilt on top of it. With much more populous cities—such as imperial Rome, or London and other major European cities, particularly after the Industrial Revolution—the vast increases in pollution, discards, and effluvia brought waste problems to the olfactory attention of elites, but these were usually tackled with mixed results at best (Melosi 2005; Sennett 1994; Reid 1991; cf. Mumford 1961).[15] Figures from late-nineteenth-century New

York City regarding horse-related waste alone give a sense of the massive scale of the nonhuman urban waste predicament.[16] Early modern Tokyo was an intriguing exception, a model, relatively speaking, of urban waste management on a grand scale. Canny frugality, sustainable organic materials, and extensive reuse extended to a broad range of spheres, and even human excrement was husbanded conspicuously rather than divorced from everyday life.[17] In a variety of forms, then, Tokyo—via its waste—diffused throughout much of the surrounding Japanese countryside in an early augury of the capital's present-day toxic wastescape.

The link between fetid conditions and public health problems was not proven conclusively until the 1880s, and urban life both before and after the discovery of germs was notoriously plagued by wretched living conditions and disease epidemics (Melosi 2001, 2005). But the twentieth century ushered in an age in which waste threats reached a higher order of magnitude. Due to technological and chemical advances and the vagaries of "risk" (Beck 1992), most waste that industrialized (and industrializing) countries produce can bear toxins that attack human and animal populations far from its source of production and at a level which can threaten those societies' health and very survival (Joy 2000)—a marked divergence from the more localized, manageable threats of the past. This very real menace dogging contemporary waste issues has a tendency to bring lofty theoretical discussions quickly down to earth, a quality that is not entirely unwelcome. The vast proliferation of toxins in everyday life (Casper 2003) has so wedded waste and illness in the minds and discourse of communities to the extent that traditional boundaries of hygiene and exclusion are influenced by toxic anxieties. In the shadow of environmental disasters in Bhopal and Chernobyl (see Fortun 2001; Petryna 2002), communities now openly contest the insalubrious effects of factory waste and energy pollution, and they harbor deep suspicions of government siting of waste facilities.

It is therefore not surprising that, until recently, waste was hardly a feature of national or municipal self-promotion. In contemporary societies waste has loomed as the accursed byproduct of edifying commerce.[18] Interest in recycling, or at least in being swathed in the raiments of "sustainability," has begun to transform this state of affairs for governments, corporations, communities, and other entities.

Of course, recycling and reuse have long been commonplace, persisting even in image-conscious societies such as America where resource-sustainable practices such as outdoor line-drying of laundry are widely eschewed as signs of poverty. Even in such ambivalent waste spheres, intricate circuits of reuse lead to virtual kula rings (Malinowski 1922) of informal relationships and exchange that exist alongside highly developed patterns of consumption and disposal: potted plants are distributed to friends and neighbors before a long move; unwanted crockery or furniture are passed down from elder to younger relatives; baby clothes, high chairs, bassinettes, and the like have such brief utility and considerable bulk that they often travel great distances for long periods of time through networks of young parents before returning, if ever, to their primary owners; "re-gifting" by taking unused or (in the Japanese case) sometimes completely unopened, still-wrapped presents and passing them on to others at the first appropriate opportunity; and so on. Equally important are the reasons why societies shy away from recirculating certain other objects. While accepting a secondhand mattress or futon in Europe or North America is not unusual, particularly during one's student days, some of my Tokyo informants reacted with visceral repugnance to such an idea: futons, for them, are intimate items, easily contaminated, and they simply could not imagine accepting them from strangers (as from a secondhand exchange venue) or even from neighbors. Kevin Hetherington (2004) argues that disposal is an overlooked but important stage in the "social life of things" (Appadurai 1988). If so, then the late-twentieth-century upsurge in recycling constitutes something of a process of reincarnation and reincorporation. Similar operations are at work when, through aggressive recycling, the very material substance of a "thing" becomes reprocessed into successive avatars in further production cycles.

In community Tokyo, and elsewhere in Japan, such acknowledgment of resource histories was common in the postwar period before the recent vogue of recycling and sustainable policy: in neat reciprocity, small trucks that wended their way through communities collecting newspaper and other pulp consumables distributed rolls of recycled-pulp toilet paper as recompense to donors—thereby contributing to the elimination of other (human) wastes. But other forms of waste were less easily divorced from the ontology of defilement that has shaped core elements of Japanese society throughout its history. The pollution that waste can engender—whether due to toxic emissions or as a result of germs transmitted by vermin or through contact—is materially distinct from aversion to what is socioculturally deemed "unclean," but dread of toxic risk is not always *subjectively* different than Japanese queasiness with "filthy" secondhand goods or the "social pollution" that low-status social groups or defiling acts are thought to transmit in a social context oriented toward widespread abhorrence of such blights. Indeed, the extent to which waste is normally kept segregated from spheres of daily life in contemporary Japan, as in other societies, mirrors other processes of segregation there that are, sadly, all too "Japanese." Though formal village ostracism *(mura hachibu)* is largely a thing of the past, operations of exclusion persist in a society that is at the same time group-focused, vertically configured, and susceptible to broad conformist pressures. An (at most) quasi-disenchanted postindustrial society, Japan remains a place where ideas of purity often radiate with significance. And waste, when scrutinized carefully, is bound up in a tangle of loosely associated notions that inspire gut-level reactions such as hygiene and illness (elimination of dangerous germs), angst over waste-scavenging vermin, vestiges of Shinto-Buddhist ritual pollution, and attitudes toward "untouchables" and other outcaste or marginalized groups. Such "pariahs" can include the homeless, day laborers, former lepers, and foreigners—either directly perceived as threats to purity or indirectly involved in contaminating activities. The specter of toxic pollution that threatens community health can also lead to ruthless exclusionary practices. In a tragic, but all too common, example, the hapless sufferers of atomic fallout in the aftermath of the bombings of Hiroshima and Nagasaki faced marginalization and exclusion. They were later joined by those ravaged by the methylmercury poisoning of Minamata Disease in the 1950s and 1960s and other victims of extreme toxic pollution in postwar Japan. Even debilitated survivors of the 1995 sarin nerve gas attack on the Tokyo subway became outsiders or objects of mockery, viewed as abnormal, damaged, and possibly contaminating (Murakami 2000). Similarly today, victims of toxic waste fallout in communities can find themselves shunned by fellow residents, who effectively blame them for their contaminated (and possibly contaminating) difference. These points of intersection between toxic waste and what I term "social pollution" are hardly coincidental. They are fundamental and help train a spotlight on broader operations of exclusion in a society that prefers they remain hidden.

Wastes and Nature

The etymology of the word "waste" in English conveys the term's complexity and its nuance not only in Anglophone contexts but more broadly. A "waste" originally described a tract of land deemed unsuitable for human habitation (OED 1989), but in time "waste" took on explicit moral overtones. Conceptions of nature common in Europe in the seventeenth and eighteenth centuries viewed land as a resource ready for exploitation and improvement. A waste, under this logic, indexed a plot of land that was underused, underexploited. The term "waste," then, came to impugn not only the tract of land or region—evidence of neglect—but also the custodians (owners, tenants, or other denizens) who had let the land languish in underproduction (Harris 1989; OED 1989). As Scanlan writes, waste was, thus, evidence of "not making the *best* use of something" (Scanlan 2005, 22, original emphasis). This was a notion bound up in Christian ideas of nature of the day, a theme that Scanlan (ibid., 24) explores elegantly in his discussion of Locke, Calvin, and other period thinkers: "There appeared to be a general concern, especially within Christian thinking, that the fruits of nature are not simply given as a consequence of some existing state of abundance. In fact the belief is quite the contrary; that there is no such abundance—there is only waste—and without the vigilance of God expressed through his stewards on earth," the order of nature would revert to disorder, chaos. Of course, less spiritually inspired motives influenced use of land. Taussig (2003, 11) reminds us that wasteland has long been a source of fuel and even food for the impoverished—in Ireland, for instance, bogland "existed where the economics were not favourable to drainage or land 'improvement,' the bog remaining as a sign of defeat for advancing capitalism spearheaded by English landlords," and the poverty of early modern Irish peasants, for example, pushed them to make use of this wasteland that others viewed as unpromising or unworthy of exploitation.

This notion of improvement through human endeavor is, intriguingly, one that strays close to connotations of nature and waste in Japanese history, extending even into contemporary attitudes there. Appreciation and idealization of nature in Japanese aesthetics was paralleled by a keen interest in exploiting the land. Many scrolls of serene natural poetry and landscape art notwithstanding, Japanese have, for centuries, practiced irrigation, deforestation, and reclamation despite claims of unique nature-focused reverence to the contrary (e.g., Huddle and Reich 1987; Kalland and Asquith 1997; McCormack 1996, 2002; Kerr 2001). This was, indeed, a moral imperative. Confucianism-inspired dictates promoted aggressive, edifying exploitation of natural resources (Morris-Suzuki 1991), which led to problems of deforestation on the "green archipelago" that only the iron fist of the shogunate could control (Totman 1989). At the same time many farmers and other subalterns were long kept at subsistence (or near-subsistence) levels due to punitive taxation of rice yields, and under such austere conditions, these Japanese were forced to avoid waste wherever possible (e.g., Duus 1976; Sippel 1998; J. W. White 1988). Core virtues of Japaneseness—cooperation, harmony, conformism, frugality—being located by many nativist ideologues in Japanese society's muddy crucible of village rice paddy cultivation, lofty idealizations of nature could not obstruct the parallel impulse to create yield out of the land.[19] Significantly, improvement was central to the experience of "nature" in the rarefied precincts of traditional aesthetics, as well. Unmediated nature fell short of the cultural ideal; instead, the helping hand of the artist-practitioner was necessary to let nature reach its full potential. The Japanese garden differed from the "wilds" in that its charms were coaxed out and cultivated over time (Hendry 1997). While in German or American idealizations of nature, remote tracts of wilderness represent nature's purest, most praiseworthy form (Schama 1995; Cronon 1996; Williams 1973), Japanese viewed the wild vastnesses[20] of the uninhabited mountains or the sea as the abode of the gods and therefore perilous in the extreme.

This tension between waste and idealized "nature," and their interplay, is embedded in the topography of Tokyo itself. Indeed, Tokyo seems peculiarly apt for a study of the social imprint of waste on a nature-focused society. The capital was, to start with, founded on waste—that is, sited on territory that was largely mosquito-infested swamp and marshland and unproductive, aside from a small castle town and some village areas. The awesome transformation, begun in the late sixteenth century by shogun Tokugawa Ieyasu, of this uneven wasteland into Edo (as Tokyo was formerly known) was a daunting odyssey of dredging and reclamation in addition to the more typical labor of city-building. And ever since, the capital has struggled with the burden of waste from its hordes of denizens living in close quarters. Edo was the world's most populous city in 1720 (Cybriwsky 1998). Ironically, some historians argue that it was precisely the city's success in controlling miasmatic conditions—particularly those of waterborne disease spawned by human waste disposal[21]—that led to this society's relatively steep trajectory of development vis-à-vis mainland Asia during this period. Expressive culture, both high and low, was one beneficiary of this age of salubrity, security, and prosperity. Edo saw an efflorescence of artistic development, much of it emerging from the "floating world" of the city's pleasure quarters and surrounding milieux and funded largely by the wealthy but low-status merchant class. In both courtly and bawdy settings, the theme of "nature" emerged as aesthetic coin of the realm, invoked by poets, geisha, musicians, and painters (amateur and professional) as the ur-motif in expressions of beauty, refinement, and urban nostalgia for the rural (see Seidensticker 1991a; Hoffman 1986; Williams 1973; Jay 1993). This widespread preoccupation with nature, which developed into a vast and intricate arena of cultural production, transformed the topography of the capital. Whole precincts of Shitamachi, the "downtown" area of the city where the floating world flourished, were literally oriented to frame views of distant Mt. Fuji and other natural-cultural features. Today the names of hills, plains, and canals ring with natural imagery, though these appellations are often their only surviving feature in an intensely developed urban environment (Berque 1997a, 1997b; Jinnai 1995; Jinnai et al. 1989).

With land scarce, Tokyo-ites disposed toward adulation of "nature" in culturally accepted forms dispensed with this reverence in creating habitable space wherever possible in the boggy ecology around them. Much of the thriving low-city Shitamachi area rose from such murky origins (Seidensticker 1991a; Cybriwsky 1998). Land reclamation continued as a development strategy throughout this period and onward, and daily waste was thrown in with other material—such as small mountains leveled for fill— to bury marshland or reclaim the "waste" of the bay. The Japanese term for reclamation, *umetate,* is a compound of the verbs "to bury" and "to build." In recent decades the use of household and industrial waste in land reclamation endures, not only providing fill for a project but furnishing a ready receptacle for the waste. Waste disposal, then, becomes a major rationale for aggressively pursued coastal construction projects in or around metropolitan areas (McCormack 1996, 2002; Kerr 2001; Huddle and Reich 1987; Broadbent 1998; Cybriwsky 1998; Seidensticker 1991a, 1991b).[22] These projects, which are notorious for their shaky economic fundamentals, also provide unpromising structural foundations: reclaimed land is subject to subsidence (as witnessed by Japan's sinking Kansai International Airport, built on an artificial isle of waste) and provides an unstable ground in earthquakes, which "lay waste" to structures on reclaimed land with greater ferocity than those on sturdier foundations. Such topographical interplay between "nature" and waste makes the capital a provocative backdrop to my study.

The Chapters

Waste is often conceived of in sweeping terms, particularly in the context of urban waste management in sprawling metropolises, yet its detri-

mental effects are experienced locally, and toxic pollution nearly always
has specific sources of generation and concentration. Chapter 2 delves
into a toxic pollution protest against a waste facility in Western Tokyo.
A key node in Tokyo's waste disposal network, this facility was first
embraced as a boon by much of the community. Just after commence-
ment of operations, however, a substantial percentage of residents surrounding the facility reported puzzling and at times acutely debilitating
symptoms. Ensuing clashes over the source and effects of toxic fallout
polarized the community. In this noxious, inharmonious atmosphere,
there operated dual processes of exclusion. On the one hand, afflicted
residents sealed themselves and their families off in their homes, seques-
tering themselves as much as possible from contamination. And on the
other hand, residents who steadfastly denied the existence of toxic waste
in the community shunned afflicted protesters and their allies, relegat-
ing them to pariah status. In addition, sufferers were ostracized by the
wider community due to their ill and possibly contaminating state. This
included social contamination, for in keeping with local sensibilities,
many Japanese avoid people who create a fuss, no matter what the reasons. The furor, which pitted the Tokyo bureaucracy against a grassroots
protest group, with each side wielding competing sets of "scientific" evidence, brought contemporary attitudes surrounding health, illness, and
purity to the fore and created a core group of residents who began to
embrace the terms of the global environmentalist discourse in making
sense of and challenging their surroundings.

For the majority of Tokyo communities, however, troubles surrounding toxic pollution and other wasteborne malaises were communicated by the media, the vehicle by which broader anxieties over illness
and contamination were often interpreted in local settings. Chapter 3
introduces my other primary fieldsite, a community in the same Tokyo
ward as the protest site but located some distance away, which endured
a barrage of media reports regarding all manner of ills, including waste's
toxic fallout. The chapter first visits a marginal forest area outside the
city limits, Kunugiyama, that was basically taken over by unscrupulous
private incinerator operators. The chapter then scrutinizes the tracery
of rumor, gossip, and hearsay that circulated throughout my fieldsite
during the period of my research, in particular the uproar over a nationwide toxic pollution scare that had its origins in blighted Kunugiyama.
Though residents in my fieldsite had, for the most part, little to complain about regarding toxic pollution and the problems of waste, their
perspective balances that of the far more politically charged protest site

and gives a glimpse of the ways in which mediated knowledge influences environmental engagement over a specific social terrain.

The preceding chapters suggest the extent to which environmental engagement is, at root, an intricate and socioculturally nuanced process with interplay between the abstract and the concrete. Next, I analyze lofty Japanese idealizations of "Japanese nature," much invoked in both exalted and mundane settings, and then I compare these with the responses of contemporary urban residents to a specific socioenvironmental problem: a vermin epidemic. Chapter 4 scrutinizes rhetoric and action regarding nature and Japaneseness that surfaced in community festivals and in the leisure sphere, and which are essential to understanding contemporary environmental engagement. Chapter 5 describes how *karasu*, swooping black jungle crows with rapacious appetites and jarring cries, became a part of Tokyo's environment that most residents would have preferred to edit out of the ecology. Significantly, the population of *karasu* had ballooned largely in proportion to the marked increase of household waste in Japan since the high-speed-growth eighties, for *karasu* fed off of domestic waste left out for community collection. These tens of thousands of scavenging crows were therefore both a nuisance growing out of Tokyo's waste predicament and an indication of how easily Japanese constructions of nature and order could be subverted by elements of the archipelago's own emergent ecology. Though at first glance a small-scale community dilemma, Tokyo's own government was forced to mobilize to control the deteriorating situation. Because of the crows' adaptable ingenuity, their occasional attacks on citizens, and the public outcry regarding this conspicuous nuisance, Tokyo's colorful current governor famously declared "war" on these vermin in 2001. This case therefore allows us to examine the broad ramifications of a specific waste quandary and a very Japanese attempt to enforce "sustainability" through scorched-earth tactics. The problem at the same time stimulated homegrown discourse on the topic of wasteful consumption and debates over the human role in creating and exacerbating the *karasu* nuisance.

The normative spin put on emblematically Japanese traits by conservative currents in Japanese society leaves out many social groups that do not meet rigid membership criteria. Chapter 6 examines operations of social exclusion in Japan—processes rarely acknowledged but very much present in this contested sphere. In a society where status and group membership are paramount, social exclusion looms as a base, ugly counterpoint to the exalted meritocratic ideals of the Japanese postwar system. The long-shunned "untouchable" *burakumin* out-caste and other cruelly marginalized groups (such as a large population of all-but-"native" Koreans, born in Japan to Japan-born parents but denied full citizenship) are well known to observers of Japan and bear many similarities to disenfranchised groups in other societies, including their proximity to waste and toxic conditions. But less well known are the cases of "mainstream" Japanese who fall ill, due to environmental illness or through contagion, and then find themselves ostracized by the society that earlier embraced them. Though numerous victims of this exclusionary process exist in contemporary Japan, I focus particular attention on certain victims of toxic pollution and the ways in which they responded to their new status. The "social pollution" perceived by Japanese groups bears numerous similarities with environmental pollution and other forms of ritual pollution that linger on from Japan's spiritual tradition, and so the chapter places this ostracism in the wider context of exclusionary thought, and action, in Japan.

Present-day Tokyo, polluted and socially fragmented, contrasts markedly with idealized constructs of traditional community life, bound together with close family ties and imbued with warmth and fertility. Chapter 7 scrutinizes the slippage between these community ideals and the comparatively disconnected urban lives of most Tokyo residents. In particular it looks at how toxic waste issues have ignited anxieties over fertility and family health. Fertility is not just a family problem but a concern of the state, so this chapter also traces Japanese pronatalism from pre–World War II colonialist designs, which intensified during wartime, through to the high-speed growth period and its own particular demands on state priorities and state control. Japan's present-day convergence of relative economic difficulty and a much publicized birth-rate decline (coupled with an aging population) make fertility, procreation, and infant health highly charged political issues. As a consequence, media reports on how toxic waste attacks the human reproductive system and impedes fetal and child development—even genital development—created pervasive dread, particularly when the waste was associated with dioxins, which are internationally notorious toxic substances. Such threats spurred a series of government measures that tackled both toxic waste and public opposition to state waste management strategies. The chapter looks at specific anxieties surrounding fertility, family, and virility in my fieldsites, including the "geography of blame"—to adapt Farmer's (1992) meaning to the microscale of residential life—that sprung from birth defects and miscarriages suffered in proximity to waste facilities. It also examines how waste issues impinge upon Japan's visions of its future.

Chapter 8 examines intensive Japanese efforts to foster sustainable lifestyles and waste processes in Japan. The chapter begins by critiquing the very notion of "sustainability"—a shifting and hopelessly imprecise notion that can mean vastly different things to divergent camps—and identifies areas of sustainability with which Japan has long been uncooperative (such as whaling), which lays bare the sociocultural underpinnings of some environmental policy. Of course, it is ironic that Japan's capital was for hundreds of years a model of sustainability, with extensive symbiosis between the urban precincts and outlying agricultural areas and significant organic waste-processing in what became the world's largest city. Against this historical backdrop, the chapter turns to the government's nascent contemporary embrace of (increasingly) responsible toxic waste management and its efforts to combat "waste" of resources through recycling and energy conservation, the latter placed in the context of Japan's historical concern over resource scarcity in the archipelago. In a society oriented to a great extent toward mass consumption, emerging social currents of frugality and eco-focused behavior from the grassroots materialized to make waste and environmental engagement resonant social issues. This hints at significant changes occurring throughout Japanese society linked to waste. With state, metropolitan, and private initiatives aimed at extensive use of waste products as sources of energy and for material (re)production, Tokyo may, indeed, be headed—a long way down the road—toward an unlikely future as a world capital of sustainable resource management, with possible important ramifications for other, more social nuances of "waste" and "social pollution" aired throughout the volume.

The book concludes by interpreting this specific case of Japan with reference to larger environmental questions, using the volume's findings as a means toward understanding the rather underexplored contextual nature of environmental engagement and the important role that environmental anthropology can play in its elucidation.

CHAPTER 2

Perils of Proximity
An Invisible Scourge

One night in May 1996, Tsubō Yūko awoke unable to breathe. Frightened and basically incapacitated, she managed to find help until the ambulance arrived. But even though Tsubō-san—a plucky, diminutive woman then in her sixties—was able, under care, to recover somewhat, she felt a strong sense of foreboding, for this was only the most acute episode of a condition that had been worsening for about a month. "It was like a bomb had exploded in the middle of my life. Until you've felt [such a thing], you can't imagine what it's like." In the aftermath of that explosion, nothing in her life was any longer the same.

Her puzzling affliction had first materialized during a visit to Izawa's newly constructed green space, tailored to cloak a subterranean waste facility: "When I went for a stroll through the Forest Park the day it opened, I could feel something was wrong. My eyes got really irritated and they started watering. The next morning, there were these thin rashes on my cheeks where the tears had streamed down my face.... Talk about a bad omen!" She later began coughing violently, with intense pain in her throat. Her saliva developed a bitter taste and, because of swelling in her esophagus, she frequently found she could no longer swallow it down. Yet this, unfortunately for Tsubō-san, was only the beginning. Her leg muscles cramped severely and, when she went to a medical clinic to investigate her condition, her blood pressure had skyrocketed to 195, far above her normal readings. Her thinking became languid and hazy, as if she were floating her way through the days. She felt that, if anything, symptoms like labored breathing and extraordinary chest pressure helped to keep her from drifting away entirely.

The first indication that Tsubō-san's afflictions might be the result of a scourge within her community came when she traveled to a clinic outside the area, and her cough and sore throat and the sting in her eyes disappeared. Upon returning, symptoms rematerialized. At the very least, it was clear to Tsubō-san that her home, even with the windows kept tightly shut, was far from a refuge. Indeed, she often discerned a strong chemical stench, like paint thinner, when she was indoors, and found no respite there. Stress and nerve-related irregularities became so acute that she found her limbs beginning to shake. "My hands trembled like leaves on a tree," she confessed. "It got to where I couldn't even keep a simple diary [for recording symptoms] anymore." Her condition reached its nadir from the life she now only dimly remembered when she could no longer breathe, and the ambulance came to take her away. In the hospital, any contact with clothes or utensils from her Izawa home caused severe reactions, so through friends and family she was forced to buy daily necessities anew, from stores located outside her community.

Tsubō-san was only the first resident afflicted with what came to be called Azuma Disease.[1] Others living nearby soon began to find commonalities in how their bodies were responding to their newly transformed surroundings. A sextagenarian housewife named Shinoda-san had reached the point where felt like she was going to die. For two months she had been unable to sleep, save fitfully. Her eyes burned. She had severe skin rashes and a hacking cough. And she suffered from a migraine that made her head feel as if it would split open like a melon. To make matters worse, before long her breathing became difficult, and she had pain shooting down her leg. It became all but impossible for her to walk. One day it got to be too much and she, too, was forced to summon paramedics to take her to the hospital. Wincing, she recollects the agonizing period before she was forced to call for help:

> Back then the air was truly awful. You couldn't believe it.... At first we would go outside and think, "That's strange, the air seems bad, doesn't it" (okashinā, okashinā, kūki ga warui ja nai kashira). But then it just got worse.... I couldn't eat, I couldn't drink the water, I couldn't take a bath, I couldn't sleep a wink.... I would cough and the phlegm (tan) that came up smelled terrible. I couldn't believe that something that smelled so bad could come out of my own body.
>
> Around that time, I went out to visit my daughters, and [where they lived] the food tasted good, the water tasted good, the air was good. But when I got back home, everything was just the same, and my condition kept getting worse.

She was eventually diagnosed with blood poisoning *(haiketsushō)* and extreme respiratory problems, in addition to her other complaints. Before long, Shinoda-san was able to discern similarities between Tsubō-san's plight and her own.

Awaki-san, a middle-aged woman who lived near Tsubō-san and Shinoda-san, also had a difficult cough, pain in her eyes, skin problems, headaches, mental fogginess, and trouble breathing. Her finger joints had begun to swell up abnormally. Eager to give me visual proof of her plight during an interview, she held out her hands: the sections of finger between joints were, to be sure, reddened and puffed up. Perhaps disappointed by my somewhat laconic reaction, however, she produced more dramatic photographic evidence: on sheets of colored copy paper, she had created a photomontage, from her own pictures, of violent skin rashes, strange protuberances, and other irregularities that had appeared on her relatives' bodies since the waste facility had commenced operations. She then went on: "It's too embarrassing to show you, but I have swelling all over as well. Lymph nodes.... You wouldn't believe what it's like." Hesitating, and apparently reconsidering, Awaki-san then produced another page with a photograph of blue-veined breasts (hers) with lumps and discolorations. Though the political stakes involved in showing an ethnographic researcher striking images (that could lead to wider coverage of the protesters' plight) cannot be ignored, the gesture made it seem as if, in the wake of this toxic scourge, all shame had been lost (see Edelstein 1988).[2]

Like the others, Awaki-san noticed that many of her symptoms would subside when she left the community. But it was only over time, and upon finding out more about their neighbors' own reactions, that afflicted members of the community developed strong suspicions that the new waste transfer facility was the source of their troubles.

The Calm Before

Prior to the protest Izawa[3] had a cordial, inclusive serenity that made it an unlikely setting for a bitter political struggle. Intensely grounding elements of community life there, such as knowing most people by face and sometimes by name in local stores, navigating effortlessly through community byways, and knowledge of layers of biographical resonance in, say, an ordinary-looking house on the corner, made living in this urban community not so different from the easy rhythms of life in a rural village. Residents were often so snugly trussed up in the social cobweb of

the community that they could hardly avoid long sequences of social
courtesies every time they creaked down the road on an old bicycle or
crossed the train tracks to stroll down to the market—to the extent that,
if in a rush, some took to looking away, becoming preoccupied with their
mobile phone, or simply begging forgiveness for dispensing with social
niceties. The local shopping artery was a hive of activity (though not as
thriving as during the Bubble years, a point frequently raised with some
nostalgia), and merchants mixed ample chatter into their daily social
transactions with local customers.

Izawa was therefore a relatively tranquil community. Yet due to
exigencies of waste policy, quiet Izawa became a key flow-channel in
the macrofunctioning of Tokyo. Tokyo is, like all cities, an aggregation
of flows. Along with the movements of capital, labor, and culture more
commonly focused upon in analyses of "global" cities (Sassen 1991; Kirby
2009b), extensive flows of waste and pollutants (and linked environmental illness) in Tokyo expose an oft-unspoken, miasmatic underbelly of
the global city phenomenon. Some of these waves of waste and pollution
are experienced in ambient forms throughout the megalopolis. But to
the extent that toxic waste always has specific sources or sites of concentration, particularly at processing and disposal sites, hapless particular
communities throughout Tokyo have found themselves disproportion-
ately vulnerable to unexpectedly high levels of toxic pollutants. Given
the pressures of waste collection throughout the megalopolis, the financial and logistical efficiencies that Izawa's facility offered—convenient
access and the means to compress waste bound for space-challenged
disposal sites—made the Izawa facility an important node in the capital's
waste network, with thousands of truckloads of waste flowing through
Izawa annually.

Izawa thus discovered itself an unwitting host to particularly noxious and profuse toxic residues that subjected a considerable proportion
of residents to symptoms ranging from merely frustrating to debilitating
and grave.

Anatomy of a "Trouble Facility"

When Azuma Ward agreed to allow a metropolitan waste facility to
be built within its boundaries, ward and metropolitan authorities hit
upon Izawa as a highly favorable location. Nestled along the border of
the ward and intersected by a major thoroughfare heading out of the
capital, Izawa was convenient both in terms of access and egress and in

terms of political and socioeconomic concerns: that is, being contiguous with another ward, the site reduced the potential level of exhaust exposure of Azuma residents. Furthermore, Izawa was palpably less affluent than many communities in Azuma Ward and hence in possession of less political capital. But unlike waste incineration complexes and other "trouble facilities" *(meiwaku shisetsu)* like nuclear plants, all of which had sparked heated opposition in Japanese communities after the more acquiescent early decades of the postwar period (Aldrich 2005, 2008; Lesbirel 1998; Broadbent 1998; McKean 1981; Huddle and Reich 1987), the proposed facility was assumed to be bereft of the kinds of toxic problems that typically alienate host communities and thus doom some waste-planning strategies. For the complex was to be only a waste transfer facility *(gomi chūkeijo,* sometimes also pronounced *chūkeisho),* a way station that would allow waste to be removed from smaller trucks able to navigate the narrow byways of the capital, then compressed with extreme force by huge compactors, and finally placed in larger container trucks that could deliver the waste more economically to landfill sites. Since none of this waste was to be incinerated, the *chūkeijo* was billed as a benign facility that, far from harming the community, would bring state-funded amenities to Izawa free of the usual risks.

Designers opted not to follow the example of another waste complex in Western Tokyo, the "cleansing facility" *(seisō kōjō,* a common euphemism for incinerator) at Takaido, which has a half-kilometer-long subterranean passage connecting the facility to a nearby highway to reduce unpleasant olfactory and noise emissions into the residential community there. The Tokyo waste bureau[4] chose to make the facility relatively unobtrusive otherwise, however—in ways that suited its ends. For example, instead of providing an elaborate ventilation apparatus, like the 480-foot smokestack at Takaido, the waste bureau decided on a far more modest exhaust tower that attempted to disguise its purpose.[5] One could easily visit the "Forest Park" atop the facility and remain visually unaware of the *chūkeijo*'s presence. Incessant truck movement through the community did not, however, go unnoticed. Izawa residents complained of putrid odors emanating from the trucks and, reportedly, from the facility—this despite the fact that the *chūkeijo* was designed as an underground complex four floors deep to reduce noise pollution and stench.

While the subterranean placement and other features of the *chūkeijo* helped "sanitize," to a certain extent, the presence of waste in close proximity, numerous residents asserted that even before the advent of toxic consciousness there, they were uncomfortable with the idea of waste from so many unknown Tokyo households moving through the space of the community. A young female office worker summed this attitude up by saying, "It feels a little weird *(chotto kimochi warui ne).*" (Especially in Chapter 6, I explain Japanese distancing protocols designed to segregate "pollution," including "social pollution" that, through their collapsing in Azuma Ward, made this discomfort with proximity more acute.) Some of those who rejected the claims underpinning environmental protests in Izawa admitted that, incentives aside, they would have preferred the facility elsewhere, though most who did were quick to add that this preference had nothing to do with any perceived toxic threat, which they challenged.[6] Intentionally generous incentives that the Tokyo Metropolitan Government dangled in front of community leaders for siting the facility in Izawa eventually closed the deal, and these went a long way toward alleviating unease with any waste-related associations of pollution, broadly defined, for many in the community.

The *chūkeijo* was burrowed beneath parkland. The government began negotiating for development of a plot of wooded, relatively "wild" land in the center of the community. This land once included a small state institute for machine technology and had become a rare trove of verdure in an otherwise developed residential area. Government authorities proposed to relandscape the wooded land, making it conform more closely with notions of tidiness and control that epitomize contemporary Japanese ideals of nature (Berque 1997b; Asquith and Kalland 1997; Kerr 2001; Kirby 2004) and creating a far more tailored public resource than the "neglected" wooded expanse had been. In addition to more extensive grassy open spaces and severe extirpation of undergrowth, the plan for the "Forest Park," as it came to be called, was to add a sports field, a rock garden, a playground, flowers, and a recreation center, all ringed by cherry trees (the last dendrological feature being virtually *de rigueur* in contemporary Japanese public landscaping). Furthermore, the ramped truck entrance to the waste transfer facility was placed behind a man-made hillock and screened behind a "natural" veil of trees and shrubbery to obscure facility operations.

Most of the community was, in the end, won over by these incentives and lifestyle concessions. Yet a minority of residents was opposed to transforming the relatively sylvan tract. Many owners of houses along the street facing the proposed Forest Park were wary of the plan to transform the land, which had been a verdant backdrop beyond their stoops and windows for many years in some cases. And when, in the end, the city stripped the "Forest Park" of more trees than originally indicated, some residents in the area believed this had been an intentional deception and expressed irritation at the sudden change from trees to (relatively denuded) "forest."[7] The combined effect of the noise of trucks, the alleged stench of waste (whatever its source), displeasure with the capital's intervention into the community's most prized amenity, and discomfort with the many tons of Tokyo's waste streaming through the area meant that imposition of the *chūkeijo* did not go uncontested or uncriticized. It also constituted one visible dimension of transformation in the community that would contribute to the sense of dislocation that residents voiced in interviews later. The disorientation caused by toxic illness, distorting existing patterns of relation and engagement, transformed the Izawa facility from a sought-after boon to a divisive political flashpoint.

Feeling Their Environs

Early victims of what came to be called Azuma Disease triggered concern among those close to them. But only over time and through informal networks of communication such as local gossip did the ranks of the afflicted come to associate the early casualties' predicament with their own health complaints. Indeed, there was an almost epidemiological character to the spread of information regarding the mysterious blight plaguing the community. Early concern with ill health, or with ailing friends or neighbors, created pockets of anxiety that seeded discontent and grew to plague the community. Residents began to find commonalities in how they were responding to their surroundings. A number of them organized into the Group to End Azuma Disease (Azuma-Byō wo Nakusu-Kai, hereafter referred to as the GEAD). After commissioning an independent scientific study that concluded the facility was emitting dangerous toxins, the protesters began to counter what in their mind was politically tainted state evidence in order to pressure the ward and metropolitan governments to shut the facility down.

THE PROTEST GROUP

The GEAD was driven by nearly twenty core members and supporters who had either felt the symptoms of Azuma Disease acutely or were close to other less politically activist sufferers, such as their children. All were more or less community-oriented residents offended by the trespass of a waste facility belching out toxic fumes in the heart of their community. A majority of the members were women, most of them middle-aged or older. In a historically male-dominated society, this constituted one considerable strike against them, on which I elaborate below. The protestors explained this preponderance of female sufferers as resulting from the normative role of women in Japan. For, due to Japan's typical gendered segregation of labor, men most often left the community every day to work elsewhere; women, in contrast, were more likely to stay home and were therefore exposed for more of the day to the localized toxins. This also, according to protesters, explained the greater numbers of elderly Japanese among those reporting symptoms. Another linked notion explaining the relatively high number of women sufferers, voiced even by some of the afflicted themselves, was that the relatively high number of women sufferers was rooted in innate feminine sensitivity, a conclusion far from unusual in the Japanese social milieu.

There were two principal male members of the group. One of these, Kimura-san, served with Tsubō-san as one of the two main spokespersons for the group. He claimed not to suffer the symptoms of the disease but regularly peppered interviews and other conversations with meaty coughs—producing an impression of illness as well as of noble fortitude, neither of which hurt the cause in the least. He worked as a playground monitor at the local primary school located near the waste facility—whose pupils' exposure to relatively high levels of a range of dangerous toxins was later admitted even by the recalcitrant Tokyo waste bureaucracy—and also acted as a self-styled ecology instructor to the children. The other principal male member, Yamada-san, signed written statements on behalf of the GEAD and offered the services of his office in providing stationery, copying services, postage fees for sending out questionnaires and appeals to government agencies, and so on. Yamada-san in particular felt that his presence was necessary in order to lend credibility to their cause in the male-biased sphere of Japanese officialdom, as well as further heft in the face of a skeptical community. He commented, wryly, "Some men only take other men seriously." That his two children suffered from exposure to Azuma Disease remained, however, a key factor in his decision to participate.

The remaining women spanned in age from thirty-four to sixty-six years old. They varied in terms of social and economic position, some well-off, others not. But the trials of Azuma Disease and the difficulties it ushered into their lives—hostile officials, neighbors denying the very existence of the condition, acquaintances singling them out for ver-

bal abuse or social frostiness—brought these members into tight relations of trust, dependency, and support. The group met formally only
infrequently, with extraordinary sessions when the occasion demanded.
But the support network that the group engendered was sustained with
innumerable phone chats, exchanges of documents, and informal social
visits wherever possible.

The small scale and marginality of the group influenced, without a
doubt, the GEAD's constitution and dealings. As Douglas and Wildavsky
(1982, 137–40) predict with their "sectarian" group model, the Japanese
protest group remained rather egalitarian for such an entity in a culture
dominated by hierarchical considerations (Nakane 1984). And like other
environmentalist groups around the world (e.g., Berglund 1998), the
GEAD treated science with reverence and depended on the legitimacy
of this form of knowledge to pursue their case. But unlike other groups,
and diverging from Douglas and Wildavsky's model, the GEAD were
comparatively civil and far less jaded in their dealings with and discussions of their opponents (though understandably within limits). While
many activists engage in aggressive public protests, civil disobedience,
and litigation, the GEAD remained an organization of which even an eti-
quette-conscious *obā-san* (grandmother) could feel comfortable being
a member. Their campaign was enlivened by not a single demonstration, formal complaints went through established government channels,
and there was—perish the thought!—no commando attempt to sabotage
the daily operations of the waste facility. Dealings with the mass media
were not attention-grabbing, as might have been the case elsewhere, but
rather, media-savvy out of proportion to the very limited experience of
the group regarding such matters.

Strikingly—in particular when I first approached the GEAD in
1999—most members voiced an unusually optimistic level of faith in the
reception that their arguments would have in the corridors of power in
the Tokyo Metropolitan Government. Perhaps this was the clearest sign
of all that the protest group was an amateur undertaking new to the process. Outside of the intense suspicion that quickly developed between
the GEAD and the Tokyo waste bureau, the GEAD remained convinced
that once their complaint was officially made to the Tokyo government,
the merits of their case and the scientific evidence they had accumulated would be incontrovertible, leading to swift resolution of the problem. This comparatively naïve approach, a combination of vestigial trust
in the state and an outsized reverence for technoscientific knowledge,
shaped the articulations and actions of the GEAD and distinguished it

from more seasoned, cynical, and Machiavellian environmentalist campaigns in North America, Europe, and Australia.[8]

One consistent theme in conversations with these protesters was, indeed, their amateur standing. These were "ordinary" people whose lives had been stripped of their ordinariness. They continued to voice their affection for the community—at least for what it had been before the arrival of the waste facility—and stressed that they were otherwise disinclined toward creating a stir. Women sometimes invoked their amateur grandmotherly status as a mark of pride: as Tsubō-san once said, meekly, "But I'm just an *obā-san*. What can I do?" Invariably, of course, there was a hint of false modesty to such comments, for in Japan it was already clear that there was actually quite a lot that a group of *obā-san* could do when they put their minds to it.

Consider the effective citizens' environmental movements of the postwar period. After Japan had resurrected itself from the rubble of the immediate postwar period, the nation enjoyed phenomenal economic growth that has widely been dubbed an "economic miracle." Unfortunately, this dynamic time also spawned the Japanese pollution debacles of the 1950s, 1960s, and early 1970s—"the world's worst health damage from industrial pollution" up until that time (Broadbent 1998, 14). The four greatest tragedies that came to light during this period were (1) *itai-itai* disease[9] in Tōyama, resulting from decades-long discharge of toxic metals, including cadmium, into a nearby river; (2) petrochemical industrial pollution in the city of Yokkaichi, where fish became inedible and where thousands of locals were exposed to severe air pollution, resulting in acute asthma problems; (3) the notorious case of methylmercury poisoning in Minamata Bay during the 1950s and early 1960s, where hundreds were killed and thousands crippled for life; (4) and another severe case of Minamata Disease (methylmercury poisoning) that appeared in Nīgata Prefecture in 1965 (George 2001; Kalland and Asquith 1997; Broadbent 1998; Japan Environment Agency 1999a). In addition to these more publicized cases, communities all over Japan bore the environmental brunt of the nation's obsession with growth-at-any-cost.

Before long, citizens' groups rose up in protest against this environmental defilement and health degradation, culminating in an upsurge of activist politics in the early 1970s. Due to a subsequent unprecedented increase in the election of progressive mayors and governors, the ruling Liberal Democratic Party (LDP) took preemptive action by pushing fourteen antipollution bills through the Japanese Diet in 1970 (Broadbent 1998; Institute for Global Environmental Studies 1999).[10] Harsh court decisions over Minamata Disease and other major pollution victims' suits between 1971 and 1973 tarnished the industries and localities responsible, putting pressure on ministries to interpret strictly the antipollution laws newly on the books. In time, more powerful conservative LDP politicians joined in, and Japan experienced something of a turnaround in its environmental fortunes by the end of the decade.

But with fewer cases of extreme pollution in public view, environmental activist political movements lost momentum. Decades later, then, Azuma Disease in Izawa unfolded over relatively apathetic social terrain. In Izawa there remained as well a great difference between those afflicted by the blight and those who were not. In the absence of mobilizing factors such as slag heaps, putrid and smoke-filled air, or waters littered with the flotsam of dead fish, the impulse toward activism came down to whether or not one actually *felt* the danger to one's family and community. This "activism threshold," as it were, differs between societies and even between different groups within societies. In Japan, whose comparatively group-focused citizens typically prefer to avoid causing "trouble" *(toraburu)* of nearly any sort, the activism threshold remains relatively high. However, with Azuma Disease a vast interpretive gulf separated those affected by toxic symptoms and those who were not affected, due to the role of "sensitivity" that distinguished Izawa's toxic pollution protest from others like it.

Contested Symptoms

The controversy over environmental illness in Izawa was magnified greatly by the contested nature of the pollution itself. In notorious cases of extreme toxic contamination such as in Bhopal or in Chernobyl, relatively straightforward evidence of considerable human suffering can be complicated enormously by factors such as difficulties in definition and quantification of illness, political jockeying among interested parties, pandemic corruption, and inequities in access to information and resources for those most debilitated by the environmental disaster in question (Fortun 2001; Petryna 2002). In Izawa, by contrast, where toxic defilement was not—objectively speaking—nearly so catastrophic, heated infighting ensued as to whether *any* toxic pollution existed at all. While numerous residents suffered acute disabilities due to the toxins in their midst—measured by a Tokyo-based academic toxicologist originally commissioned by the Tokyo government who, after seeing his findings twisted by the government, conducted subsequent measurements for the protest group—other members of the community apparently unaffected by the pollution found it very difficult to accept protesters' claims. These differences of opinion stemmed, in great part, from the type of facility that lay beneath the Forest Park's manicured grounds.

Ward and metropolitan authorities had gone to great lengths to stress the transitory nature of the waste the Izawa facility would handle. Since this was not to be a long-term storage site, and no incineration of waste would take place there, so, the reasoning went, none of the ills associated with waste facilities would apply. In practice, however, things worked out differently.

The mystery of this facility, and source of bitter controversy surrounding Azuma Disease, was that dire symptoms of environmental illness materialized despite the fact that there was no incineration at the site. That dioxins—including the most lethal man-made substance on the planet—and other dangerous toxins are easily released into the atmosphere through incineration of plastics has long been demonstrated by scientific studies. Since Japan burns about 80 percent of its waste, airborne toxins belching out of Japanese incinerators were a health threat already familiar to members of Japanese communities, whether or not they themselves were actually exposed to dangerous levels. The topic of airborne toxins was very much "in the air," so to speak, in the mass media and in local discourse (see Chapter 3). With Azuma Disease, however, members of the GEAD claimed that, *somehow*, toxins were being released through the powerful compression of plastics by industrial-scale compactors deep within the bowels of the Izawa waste facility. GEAD members argued that the ventilation tower filtered the air leaving the facility only for odors and the like rather than more dangerous substances. As a result, toxins were being discharged directly into the air and blanketed the community. And since the facility was said to be among the first of its kind in the world, the protesters went on to claim that science had not had a chance to verify the dangers of violently compressing plastic, leaving residents vulnerable.

What made their efforts to convince others so difficult was that, according to protesters, not everyone responded to the toxins in the same manner. According to systematic GEAD studies, approximately 10 percent of the population living within 500 meters of the facility reported toxic symptoms that, in aggregate, came to be called Azuma Disease. Afflicted members of the community, presented with this puzzling inconsistency of illness, chalked it up to the vagaries of chemical hypersensitivity *(kagaku-busshitsu kabinshō)*. According to this explanation (which has a provenance in Japanese and international pathological research; see, for example, Ishikawa et al. 2003), some individuals—like Tsubō-san—with a low tolerance for toxins fell prey to them almost immediately. Others carried on with their lives as though nothing had changed in their environment, though their health still suffered, according to the claims of GEAD protesters and environmental scientists, as a result of daily exposure. Most of the rest of the population, including the 400-odd residents afflicted with varying levels of Azuma Disease, fell somewhere between these two extremes. Due to the vicissitudes of chemical sensitivity, then, a facility might have a grave effect on a fraction of the population while seeming relatively innocuous to the majority. Such was the case, protesters reasoned, in Izawa.

To be sure, theirs was a controversial claim, one that was at first mocked and then more systematically discredited by government officials and local opponents. But even in cases where more conventional arguments prevail, toxic pollution and the illnesses that it triggers are extremely difficult to prove. Since most chemicals, even such lethal carcinogens as dioxins, are statistically rather harmless in small amounts, the assessment of risk comes down to scientific analyses and shifting definitions, many of which are open to interpretation, influence, and downright sabotage. Much toxic pollution is, furthermore, imperceptible to the senses—dioxins, for example, are tasteless, odorless, and invisible to the naked eye, discernible only via specialized technoscientific instruments that laypersons cannot themselves employ. While those battling over determinations of toxic pollution are thus often limited to the realm of language (Nickum, Midori, and Tokashi 2003; Berglund 1998), depending on findings collected by specialists, the afflicted in Izawa experienced toxic damage with the subtle antennae of their bodies. Izawa lacked the litter-strewn expanses, the choked, contaminated waterways, the industrial residues, and the visible smog that could mobilize large percentages of the population. Yet for the afflicted in Izawa, their bodies *themselves* became like desecrated landscapes, a point that surfaced in their discussions below of damage to the "nature" of the Forest Park. Therefore, while many in the community remained skeptical, the protesters themselves had no doubt whatsoever that they were suffering from the ravages of toxic pollution and illness; their task was to convince others. The GEAD and its supporters, then, behaved rather like prophets whose visions remained unrecognized by the masses (see Douglas and Wildavsky 1982).

Importantly, GEAD members chose the term *kanjiru* (to feel) when describing symptoms of Azuma Disease. Use of this verb rather than some more clinical term like *kakaru* (which can be translated as "contract" or "catch") reinforced the belief—at least for the GEAD faithful—that the disease was out there and, though some might not feel it, they were affected by the toxins regardless. Moreover, "feeling" here implies a special sensitivity, an engaged subjectivity as opposed to the clinical-sounding, intransitive verbal form that "contract" takes in Japanese and the implicit associations of passivity with which the term resonates. It is worth pointing out that sensitivity and illness have long been associated with intellectual or creative genius in Japanese society. According to Japanese folk understandings of illness and the body, people are thought to possess different constitutions, or *taishitsu*, and some *taishitsu* are seen to predispose people to illness.[11] This is not a low-status, pariah condition (Ohnuki-Tierney 1984, 51–52), but one associated with authors, artists, intellectuals, and so on. Indeed, there is a certain cachet in Japan to being *shinkeishitsu*, *horyūshitsu*, or *senbyōshitsu*—all indexing a sensitive or weak condition—whereas being *jōbu*, or healthfully *mushinkei* ("no nerves or sensitivity"; ibid.), is akin to dullness (see also ibid., 61–62).[12] These nuances did much to elevate GEAD member Tsubō-san's standing among her peers. While Tsubō-san would likely have been a leading member of the group anyway, her status as the first to have "felt" Azuma Disease set her apart. Being the first, she was seen to have a unique sensitivity, akin to special, almost mystical powers. For instance, voicing a notion that I heard from many involved with GEAD activities, Shinoda-san responded to a question about chemical hypersensitivity *(kagaku-busshitsu kabinshō)* by saying, "So Tsubō-san has the highest *kabinshō* because she felt [the disease] first; but there are all these other people who feel [Azuma Disease], and that's *kabinshō*, too. Some people feel more than others." Some members seemed to be eager to demonstrate how early they "felt" the toxins plaguing the community. This reverence for primacy, or near-primacy, is widespread in hierarchical Japan. Being first, or among the first, gave a handful of GEAD protesters enhanced social status within the protest group and extra cachet at conferences or meetings where they came into contact with activists from other arenas. In essence, unusual hypersensitivity conferred something akin to (very limited and unremunerative) celebrity status—small comfort when recovering from extreme debilitation, but important nevertheless and of significance to this study.

Thus it was on such fraught, destabilized social terrain that bodily engagement of the afflicted in Izawa transformed into political engagement. Yet due to the invisible nature of the toxins, the difficulty of contesting technoscientific data compiled by the state, and the limited cohort of afflicted residents, the brunt of the blame shifted to the protesters themselves (Farmer 1992).

Blaming the Victims

I must make brief mention of the protesters' nemeses, the waste facility and its workers. Ideally, I would have been able to interview employees and supervisors extensively, which would have allowed me to conduct an in-depth analysis of the facility's side of the story. I would then have been able to assess what employees saw themselves as doing, how they conceived of environmental risk, and I would have had a better chance of finding or corroborating official and unofficial attempts to bowdlerize details of toxic pollution, disguise its effects, or paper over dissent among facility workers. I would also have been better positioned to determine whether there had been any attempt to hide sickness or fear of sickness among the employees.

But the administrators of the waste facility kept their employees on an exceedingly tight leash, making official interviews impossible. I did manage, however, both to strike up some conversations in the *chūkeijo* lobby and to get a short look around some of the facility. Though it was difficult to assess the extent to which I was merely being served the official line, my guide Koyanagi-san, a full-time employee there, was adamant that there was no scientific basis for the protesters' claim that compaction of plastic "unburnable" waste could lead to release of toxins into the air. He believed, in fact, that there was no danger whatsoever from the running of the facility. Chinks in his armor-like certitude were exposed some years later, however, when even the recalcitrant Tokyo government was forced to admit toxic emissions (see Chapter 8). When asked about the possibility of illegally dumped or collected industrial and toxic waste finding its way into the compactors, he dismissed the idea, stating that the Tokyo waste bureau was vigilant of such illegal dumping. He seemed not only defensive regarding the GEAD protest but also proud of his waste facility's role in the vast complex of the Tokyo metropolitan waste network. The power of group membership in Japan is not to be underestimated—particularly when comments are given on the clock, so to speak—yet from the positive comments of Koyanagi-san and other employees, it was clear that there were two sides to this debate.

The bulk of the Izawa establishment weighed in against the GEAD troublemakers. More than a few residents in Izawa believed that the GEAD were crackpots or simply *obā-san* hypochondriacs with nothing better to do but to invent maladies and problems for the rest of the community. From the leaders of the community down to less-engaged residents, Izawa, by and large, condemned the arguments of the GEAD, at least publicly. GEAD members certainly expected people not suffering the symptoms of Azuma Disease to be skeptical. But there were several points about which the protesters could not help but be suspicious. To begin with, they believed that some residents were pleased enough with the status quo (including the Forest Park and its gleaming recreational facilities) that they did not want to know the truth. Assuming that these included the portion of residents not suffering the effects of toxic exposure, protesters conceded that good health certainly must have played a role in their neighbors' complacency. Next, the GEAD and their allies suggested—accurately, it seemed to me—that some antagonists to the GEAD's activities were driven by baser concerns, such as protecting the value of their real estate holdings and local businesses. As self-interested landowners or merchants would swiftly calculate, a community known to suffer from toxic pollution could expect deflated property values, reduced rents, and lower business volume. In the mind of the protesters, such vested interests had strong financial incentives to try to quash rumors of Azuma Disease. Finally, in Japan as elsewhere, reputation is everything. Izawa—like my other fieldsite, Horiuchi, and, indeed, many other communities in Tokyo—was a blend of natives and newcomers (Robertson 1991; Bestor 1989). Established families might have lived in the area for generations, with strong ties to the land, including agricultural ties. Newcomers, on the other hand, moved to the area with the postwar development of Tokyo's outlying agricultural regions. Though some of these newcomers might have lived in the area for decades, there still remained a distinction between them and the more established members of the community, which was demonstrated in their participation in and influence at local *chōkai* or *jichikai* meetings. These long-time members of the community were much less likely to resist the facility or join the protest group. Furthermore, they had, arguably, a stronger vested interest in keeping Izawa free of the taint of the toxic—if only in name. Though the protesters eventually forced to flee Izawa were very sad to leave the community, they usually had less to keep them in that specific area of Tokyo than people with a strong literal and imagined connection to the land of their *furusato,* or "native place" (Robertson 1998, 110, 115). To be sure, however, the experience of seeing (or

feeling) the community turn against them went a long way toward transforming their conception of Izawa, an idea to which I return in Chapter
6 in discussing social pollution.

While the community maintained a largely monolithic resistance
to the activities of the protesters, sympathetic residents found ways to
make themselves known. Sometimes friends or neighbors found themselves in a socially untenable predicament, perhaps with a spouse who
would not countenance GEAD support, or holding an occupationally
sensitive position, or being otherwise disinclined to go against the sentiments of the broader community. (The saying "The nail that sticks out
gets hammered down" [*deru kugi ha utareru*] still resonates in contemporary Japanese communities.) These conflicted residents played a quieter role on occasion; for instance, Awaki-san recalled, "I've had plenty
of people leave an envelope with money in my mailbox with a note
saying, 'I'm very sorry, this is all I can do for now,' sometimes not even
writing their name, though I think I knew who they were each time.... I
guess I feel of two minds [about it]. I mean, it's nice to have the support. But if fewer people like that stayed silent, we would've been able to
change things around here a long time ago." Though such fissures in the
community gave those afflicted by Azuma Disease reason for hope, they
also knew that their real target remained the Ward and Tokyo Metropolitan Government, whose dictates even the skeptics would be forced
to accept. Because of that, GEAD members set out to prove the existence of the pollution. The strategies they selected and their discussions
of their plight shed light on waste politics there and, indeed, elsewhere
in Tokyo's uneven wastescape.

Finding Proof

There were two main prongs to the protesters' attack on the waste facility. One focused on the available evidence of defilement within the Forest
Park and its surrounding area, a relatively unscientific undertaking that
was, nevertheless, persuasive and had excellent propaganda value. The
other prong fought science with science, directly challenging the scientific evidence that the state had assembled to justify the *chūkeijo* operations. Izawa's afflicted and their supporters were convinced that the state
had manipulated the "facts" presented, but the GEAD waged its campaign
from a disadvantageous position, without plentiful resources or institutional backing. Both sets of evidence and their pursuit revealed important
elements of the protest and of Japanese environmental consciousness.

EVIDENTIARY NATURE

The Japanese *idea* of the Japanese love of nature is one firmly embedded in a wide range of contexts, yet it seems particularly notable in its most teeming cities. The "nature" that planners chose for the relandscaping of the Forest Park's erstwhile relatively "wild" verdure, which was warmly received by the residents, was a banal form of highly mediated greenery landscaped in accordance with mainstream urban Japanese tastes. This was not the thick, lush forestland and sometimes almost impenetrable *sasa*-grass underbrush of the Japanese hinterland—forms of nature that were simply too unruly and inconvenient for comfortable consumption by most adults in Izawa. Instead, many residents preferred the sculptured grounds, the stroll-through rock garden, and the more extensive meadow space for cherry-blossom viewing, among other amenities provided by the government.

The irony of this consumption of the Forest Park's polluted amenities was not lost on GEAD members. For many of the protesters, the Forest Park was a disturbing symbol of something badly wrong in their community. Attending leisurely picnics under falling cherry blossom petals near the very waste facility that was producing the toxins that had damaged their bodies was simply intolerable.

The protesters painted the Forest Park as a sylvan sham whose function was to disguise the toxic effects of the waste facility on the community. And they set out to document the natural degradation that they discerned there—evidence that they believed not only undermined the public relations effort regarding the park but that also proved the existence of the pollution that the state denied. They carefully photographed the profusion of dying and wilting foliage they identified near the ventilation tower and the entrance to the waste facility in particular, but which they spied all over the park. They also spotlighted the activities of park groundskeepers, who, allegedly on orders of the Tokyo waste bureau, regularly hacked away at the decay. That the clearing of dead foliage is part of any groundskeeper's job was not by itself a mitigating point for most protesters, given the nature of their grievance. Kimura-san, the playground monitor, complained, "Nature can bear a lot, but this [is] too much," and he went on to lament "the dead foliage, the stench, the decline in birds and animals." As for Awaki-san's husband, he observed (with some spousal amusement), "She inspects this community with an extremely strict eye, I can tell you."

To hear the protesters tell it, the Forest Park was the scene of an ongoing battle to keep the park looking "healthy" and "natural." My inquiries on this theme with park staff and other locals indicated that this was somewhat the case, especially early on, but might have been exaggerated along the way. Thus in parallel with the GEAD's campaign against the scientific basis for the facility, they gathered a wide range of such circumstantial evidence—even including a photograph of a dead cat found near the waste facility—to lend credence to their central argument. The protesters' careful vigil over the plant life of the Forest Park matched the daily monitoring of their own health, withering flora effectively serving as an ecological mirroring of their own bodily suffering.

In tandem with this concern over the park, some protesters accumulated photographic evidence of the devastation of their bodies, and the mingling of proof of ecological and somatic damage was telling. During the many times that Awaki-san showed me her albums stuffed with snapshots, she presented numerous photographs of alleged toxic decay in the Forest Park and then brought out photos detailing the physical toll Azuma Disease had had on her own body and those of her family members—rashes, discolorations, and strange protuberances were every bit as unnatural as the scars they discerned on the landscape of the community. Both she and others of the afflicted regularly mixed together references to their bodies and to the ecology, and it was clear that their interlinkage had translated into something of an affective blurring of their subjectivities. For not only was the Forest Park's formerly flourishing "nature" a vital part of the community, but it was also visible to all in a way that protesters' bodies, often shut up in their homes or recovering far from Izawa, were frequently not. The degradation of this "commons," as it were, made for evocative visual material in a dispute with little that could serve that purpose.

This effort, though, was not the kind of ammunition the GEAD needed in their battle with the state. It was the scientific evidence the protesters gathered that was eventually convincing enough to breach the wall of denial the Tokyo waste bureau had erected and advance the GEAD's case toward unusual public hearings that brought further legitimacy to their cause.

OUR SCIENCE VS. THEIR SCIENCE

Given the murky etiology and symptoms of Azuma Disease, as well as the skeptical, even hostile community and officialdom they faced, Izawa residents afflicted with the disease came to view scientific evidence as the front line of their protest campaign. Early on, Tsubō-san recalled, the group tried to appeal to the government by describing their ill health and worsening condition. But when this strategy produced no results—indeed, stinging from the callous treatment and latent suspicion with which their appeals were met—the protesters set about trying to document their case. They accumulated a convincing amount of data that went a long way toward supporting their claim that the facility was dangerous and that the Tokyo waste bureau had been behaving in an underhanded manner both in its running of the facility and in its dealings with the public. "Science," of course, was the protesters' most powerful weapon, and what they mustered, they brandished constantly. Tsubō-san, basically the coordinator of the GEAD's campaign and presentations, traveled seemingly everywhere with files full of statistics, graphs, maps, and environmental regulations. She stated, "Scientific proof *(kagakuteki na shōko)* is the only language that politicians understand when it comes to environmental pollution *(kōgai)*....So though we didn't have very much money, we decided we had to commission a study." The GEAD realized they could manage to fund only a limited environmental assessment.

Though modest, the atmospheric testing they commissioned—performed by Arasaki-sensei, a physicist at a university based in Azuma Ward—turned out to be very good value for the money. Conveniently, Arasaki-sensei had earlier been hired by the Tokyo waste bureau to do similar testing, and while on the public payroll, he had found a high level of toxins in the air at several sites in the community. He had given these and other data to the bureaucrats, who systematically warped them in their official reports. (The manipulation of these data is patently obvious to anyone presented with the initial measurements and the later reports of the Tokyo waste bureau, who methodically used selection, omission, and distortion to obfuscate Arasaki-sensei's unwelcome findings.) These substantiated falsehoods, and Arasaki-sensei's participation in the initial tests, gave his subsequent tests and testimony a great deal of weight when they were examined in public hearings.[13] For the GEAD, Arasaki-sensei sampled the air quality in numerous sites around Tokyo, some of them places where high levels might be expected; others were residential neighborhoods thought to be relatively free of such toxins. Arasaki-sensei's research determined exactly what the GEAD had believed from the beginning: the toxin levels for Izawa were often dramatically higher than in other residential communities tested. This was damning evidence, and most importantly it had been compiled in technoscientific terms that would make the GEAD's case difficult to refute out of hand. To supplement this evidence, the GEAD conducted a survey of residents living within 500 meters of the facility, the product of written questionnaires. The responses were filled with attestations to declining health and increases in the symptoms that had become associated with Azuma Disease. These data argued that approximately 10 percent of the population within the tested area were suffering from varying degrees of toxic exposure. The activists did the best they could with the methodology to make the study "scientific": assigning random numbers to the respondents, preserving anonymity, and ranking participants with regard to the severity of their condition based on their responses. The questionnaires and other GEAD materials, in addition, were filled with clinical medical terminology of the sort that helped imbue their efforts with biomedical validity. Poignant personal accounts were combined with the respondents' answers to specific queries and then organized for maximum effect.

The obvious need for the protesters to substantiate their claims in persuasive technoscientific terms led to a marked discursive shift in how protesters articulated their sufferings and how they depicted their surroundings (see Carrier 2004). Contemporary Japanese society is, by and large, highly predisposed to respect scientific data and technological methods. So these residents of Izawa were already operating in a discursive universe where scientific descriptions and statistical data yielded deference and approbation. But as members of the GEAD sought the legitimacy to buttress their case, individuals such as Arasaki-sensei and certain Tokyo-based toxicologists and doctors took on an aura that made their importance to the GEAD quite evident. Not all GEAD members were able to become autodidact experts in toxicology, but many became highly conversant with the figures and tables that constituted their study. Many, in addition, knew at least passably well the points in the Tokyo waste bureau's study that could be considered gross manipulations of otherwise impartial data. Much of this information came to GEAD members secondhand, but there were individuals whose judgment was deemed trustworthy, particularly those who had gone on record publicly against the Tokyo waste bureau and the waste facility. The voices of these individuals carried a great deal of weight in shaping the convictions of GEAD members. In Japan there is, indeed, simply no critique of the use of technoscience as the final arbiter in disputes surrounding Truth. Those who were steeped in scientific method (Arasaki-sensei, Yamada-san, and Tsubō-san to some extent, as well as others who chose a quieter, advisory role, such as local doctors) used techno-scientific language and data with calm certainty; those who were less comfortable with science, but who clothed themselves in its discursive raiments, came to view technoscientific evidence in an almost religious light, imbuing the data (and the practitioners) with a respect that bordered on blind reverence.

In the Izawa conflict, there was a polarization of "good" science and "bad" science, largely symmetrical to that of "our" science versus "their" science. For all the pretense of impartial measurement, this was a deeply political battle that betrayed the high stakes surrounding waste issues in Tokyo. Yet as I demonstrate, such controversy over toxic waste and toxic illness, both real and imagined, also circulated throughout communities that, unlike Izawa, were less politically polarized and perhaps more representative of environmental attitudes throughout Tokyo's wastescape.

CHAPTER 3

Mediated Anxieties
Nowhere to Hide

The vituperative protest over the troubled Izawa waste facility serves as a telling case of conflict in response to waste and toxic damage, but it is hardly representative of environmental engagement in the thousands of other communities dotting Tokyo's varied wastescape. This is not to say that waste dilemmas did not impinge upon life in these less overtly polarized settings. Residents of Horiuchi, my other primary ethnographic fieldsite, had far less in particular to complain about regarding toxic pollution and problems of waste, and yet they were exposed to a barrage of mass media that gave them ample reason for concern over their surroundings and the health of their families and community.

In Japan and elsewhere, environmental pollution is typically framed in terms of sources and proximity. While environmental threats can be regional or global in scale, they frequently flow from particular hotspots, and resulting contamination diffuses into surrounding landscapes and communities. Even far-reaching environmental calamities such as the nuclear disaster at Chernobyl in the Ukraine are conceived of in this localized manner. Despite the radioactive fallout that, over time, drifted across a number of European nations and elsewhere, most cognitive and international attention remains focused on the most contaminated areas themselves or on surrounding Ukrainian populations and terrain (see Petryna 2002). A reflection of this topocentric logical tendency is the way environmental catastrophes are named. More often than not, the geographical location of the problem becomes the shorthand for the problem itself: Chernobyl, Bhopal, Three Mile Island, Love Canal, and so on. This goes for Japan as well, of course, with Minamata Disease being only the most notorious case of a contaminated locality serving as reviled emblem for a variety of environmental illnesses. Indeed, there exists a

Japanese formula for such naming, of which sufferers from Azuma Disease, described in the previous chapter, shrewdly availed themselves. The grotesque methylmercury poisoning in Minamata became known as Minamata Disease *(minamata-byō)*, pollution problems in Kawasaki were dubbed Kawasaki Disease *(kawasaki-byō)*, and so on. GEAD members opted to use the name of the encompassing ward in which they lived, rather than the name of their small community, because by implicating the whole ward in the problem of Azuma Disease, they boosted awareness of their plight and put the ward government on the defensive, effectively making the ward part of the solution.[1]

Sometimes, however, there are environmental threats that are not necessarily "global"—such as climate change—that, nevertheless, affect large numbers of people far beyond a specific locale. These problems generally do not materialize overnight, but via mass media dissemination, waves of apprehension can mimic the dispersion of the toxic pollution in question. As consciousness of a toxic threat moves, with epidemiological swiftness, throughout a population, the suddenness of the perceived danger appears to increase the level of anxiety in a community.

This chapter traces a toxic waste threat of this kind as it developed in one hotspot and then loomed over the entire nation, mushrooming into a toxic pollution scare that shocked Japan and helped push decisive action on waste policy. The narrative begins on the periphery of the capital before moving to a community that served as the second of my two primary fieldsites. The interplay between distant areas, on the one hand, and contiguous entities under different administrative authority, on the other, points to the interpretive complexity of waste issues and the difficulty in limiting one's analytical scrutiny to "Tokyo" narrowly defined.

Japan's "Fifth Avenue" of Industrial Waste

A sylvan residential area on the margins of the capital went from relative normalcy to bitterly declaring itself the "Industrial Waste Ginza" of Japan *(sanpai-ginza)*[2] in the space of about a decade (e.g., *Asahi Shimbun* February 22, 1999). The saga of late-twentieth-century waste management in Japan is in many ways etched onto the landscape of this community, and so here is, in many ways, a fitting place to start.

Kunugiyama, the area in question, is a forest zone at the intersection of four administrative districts in Saitama Prefecture: Tokorozawa-shi, Kawagoe-shi, Sayama-shi, and Miyoshi-machi, whose boundaries fit together into an intricate jigsaw puzzle of arbitrary-looking cartographic protrusions (see map 1). For years it was a quiet, unremarkable place, known locally as a wooded haven with fresh air and wild mushrooms, birds, and small animals. But by the 1990s it had become home to sixteen low-standard incinerators that were clustered within a 500-meter radius, with a further forty-two incinerators operating nearby. Tokorozawa is a Tokyo "bedtown" community with an easy commute to the capital. The city also has a good deal of agricultural land, and the area is reasonably well known for its tea and spinach. Such crops were cultivated in fields located sometimes very close to this concentration of waste incinera-tors. Perhaps more importantly, prevailing breezes from the incinerators often brought toxic smoke downwind to the Tokorozawa fields and resi-dential areas. Dioxins, including the world's most lethal man-made sub-stance (tetrachlorodibenzo-paradioxin, or TCDD), are produced during incineration of plastics, an abundant form of household, commercial,

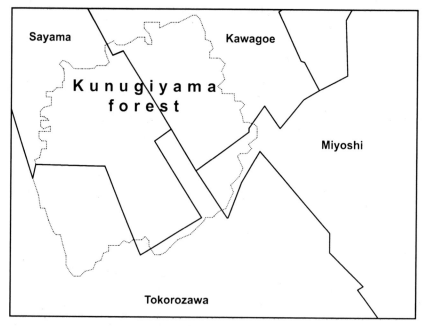

MAP 1   Kunugiyama, a forest zone at the intersection of four administrative districts just outside of Tokyo. Partly because no one city was responsible for regulating the whole area, the for-est became infested with low-standard incinerators whose smoke-stacks choked the woods and surroundings with toxic emissions.

and industrial waste. These poisons can, like other toxins, find their way via smoke-contaminated soil into the crops produced there (Institute of Medicine of the National Academies 2003; World Health Organization 1999; Institute for Global Environmental Studies 1999).

Many of the same qualities that attracted residents to Kunugiyama attracted waste disposal entrepreneurs as well. Tokorozawa is located just one exit from Tokyo on the Kanetsu Expressway, an easy drive for garbage-laden trucks. And forested Kunugiyama is located a mere fifteen minutes' drive from this interchange. Tokorozawa was, therefore, en route to final disposal sites further north/northwest—though hours away in some cases—in Gunma, Nagano, Fukushima, and Aomori Prefectures, as well as to the Hokuriku region. If the final disposal site was in the Tokyo area, Kunugiyama effectively allowed waste handlers and others to step over the capital's official boundary to avoid Tokyo's stricter waste processing standards. Much of the waste, furthermore, simply remained in the Tokorozawa area in available disposal sites, whether sanctioned or not.

Significantly, the forest zone was remote from public view, which cloaked illegal disposal methods (see figure 5) and other unsavory activi-

FIGURE 5   Detail of a sign warning against illegal dumping in Kunugiyama: "Keep our city clean."

ties. (This hooded, betwixt-and-between quality eventually ended up attracting the odd love hotel to the neighborhood as well.) That the area was under the jurisdiction of no less than four separate administrative entities also provided useful cover. Officials from one city were unlikely to monitor activities over the border, and prefectural and national laws legislated (with considerable laxity) only to limit the emissions of single incinerators, not dense concentrations of incinerators. As a result, waste companies operated here in a regulatory grey area that was as profitable as it was opaque.

While opposition from local community groups became sustained and vociferous, little changed in the area until Tokorozawa, and waste incineration generally, became embroiled in a nationwide furor over toxic pollution.

News Synopsis

On the evening of 1 February 1999, TV-Asahi's "News Station"—Japan's most popular news program at the time—featured a report on dioxin pollution in Saitama Prefecture's Tokorozawa.[3]

Citing independent sources, News Station's crusading anchor, former comedian Kume Hiroshi, reported that vegetables grown in the city of Tokorozawa had been found to contain a concentration of dioxins far higher than considered healthy by the government. This was by no means News Station's first foray into controversial territory: the program had been more or less singlehandedly responsible for dethroning corrupt LDP kingmaker Kanemaru Shin in 1992 and 1993 with a relentless nightly bombardment of broadcast criticisms of his transgressions. The latest television news report exposed the fact that dioxin-laden smoke from loosely regulated waste incinerators can become embedded in the soil and hence in the vegetables grown nearby—a shock to most Japanese, who generally thought incinerators were a danger to humans only when in the vicinity. As the Saitama region was known for its spinach, the report had immediate economic repercussions. Within days, major supermarket chains across the country had withdrawn all Saitama vegetables from their shelves, and prices had plummeted nationally.

Yet soon after the broadcast, journalists and other commentators began to cast doubt on the validity of News Station's source, which turned out to be a private research institute commissioned by the program. Eight days later, under duress, News Station aired a retraction of sorts over their choice of the word "vegetable" *(yasai)* over "leafy produce" *(happamono)*. The program was forced to admit that while the Saitama vegetables tested contained some of the toxins, in fact the reported high level of dioxins was found only in tea.[4] This and subsequent reports were seized upon by TV-Asahi's competitors and ricocheted around the domestic news media for a number of weeks.

Most journalistic stories have a distinctive news cycle: a news organ will take a negative perspective on a topic, say, and then several times over the subsequent days or weeks present the obverse perspective in a rhythm dictated largely by the perceived currency of the story and the instincts of the editors (see Bourdieu 1998). Depending on the luridness and complexity of the story, it may support numerous such flips in perspective (usually from dramatized risk to relative security) that keep the story alive and manipulate the level of anxiety, or "interest," the story can generate.

The TV-Asahi/Tokorozawa dioxin scare, it turns out, was well suited to such fugued, contrapuntal coverage. The shock of the initial story (contaminated vegetables being sold and consumed all over Japan) led to reports of Saitama Prefecture vegetables being boycotted by supermarket chains. The following week the Tokorozawa agricultural cooperative presented findings from their previous study showing vegetables to have relatively safe dioxin concentrations (e.g., *Asahi Shimbun* 10 February 1999). This prompted TV-Asahi to recant on some of the language of its initial report and, in turn, called into question the validity of their coverage. Yet amid the fallout surrounding the program's apparently intentional distortions, newspapers and television stations also began to expose the wretched quality of the air in Tokorozawa, with its pall of smoke from the many incinerators there. Tales of residents' struggles with dioxin and other pollution in the area mingled with reports from joint-ministerial government surveys of Tokorozawa's vegetables, soil, and air that proclaimed the area's agricultural produce safe for consumption. Along the way, as Japanese consumers and supermarkets hesitantly returned to buying and selling Tokorozawa produce, newspapers and television programs broadcast intermittent reports of the head of TV-Asahi apologizing to Tokorozawa farmers and explaining News Station's handling of the story in front of a critical Japanese parliament.

Aggressive news organs thrive, in many ways, on manipulating the degree of anxiety and uncertainty in the lives of its consumers. They convey images, data, and perspectives that can erode people's trust in social institutions and bring risk consciousness to the fore. With the Tokorozawa dioxin scare, this objective was achieved, albeit via a circu-

itous route, and not without TV-Asahi experiencing considerable blow-
back from the story it detonated. It is not clear, in fact, that News Station
ever truly recovered from the incident. Japan's most popular news program did stay on the air for several years longer, but Kume Hiroshi, the
charismatic anchor, left to take an unusually long vacation just after the
dioxin furor died down later in 1999, returning to work (sporting a new
beard) in early 2000. By 2004, the show was off the air. Coincidentally,
this was the same year that TV-Asahi paid Saitama farmers 10 million
yen in settlement for the toxic broadcasts.

The fundamental difference between this sensational toxic pollution media drama and other prominent ones before it was, literally, how
close it hit to home. With most cases of environmental pollution, a specific zone is labeled contaminated and a particular facility or company
or act of god is deemed responsible, allowing recipients of mass-media
reports to disengage somewhat from the story. With distance comes
detachment. By contrast, the contaminated zone of this 1999 pollution
incident ballooned to press up against the very boundaries of Japan itself.
The Tokorozawa dioxin case left open the possibility that nearly anyone,
anywhere in Japan, could have ingested contaminated produce and that
this might have been just a fraction of their total exposure over years
or decades. For example, two years earlier, the operations of a municipal waste incinerator in Nose-chō on the outskirts of Osaka resulted
in Japan's worst known case of dioxin contamination. But the case was
extremely localized—residents living around this Osaka incinerator
were, to be sure, appalled by the presence of toxins in their midst,[5] but
readers and viewers elsewhere in Japan could easily decouple themselves
from what was a mere isolated incident. With the Tokorozawa dioxin
scare, however, toxin-tainted vegetables had allegedly been transported
from a distant, isolated, contaminated site straight to everyone's local
grocery stand and dinner table. The combination of massive potential
dioxin exposure and even more massive media exposure proved to be
volatile indeed. Despite the discredited allegations of the first TV-Asahi
reports, initial shock and outrage toward the story, as well as its prolonged character, ensured that further scrutiny would be brought to bear
both on conditions in Tokorozawa and on dioxins generally. It was during these months of scrutiny and media deluge that the threat of dioxins
began to sink into wider consciousness.

The impact of the dioxin threat and awareness of Tokorozawa's specific plight were unmistakable in Horiuchi, a Tokyo community located
about fifteen miles away from the "Industrial Waste Ginza," and in the

same ward as Izawa. I focus on Horiuchi to convey the pulsations of anxiety and misunderstanding that rippled throughout Tokyo's wastescape while I was conducting research there.

Cobwebs of "Betweenness"

For more than six months before the Tokorozawa dioxin scare, I'd been based in Horiuchi, in Western Tokyo. My ethnographic research explored the gulf between the profusion of Japanese rhetoric over "nature" and Japanese society's clear history of postwar environmental neglect. Yet, inevitably, interviews with local informants dwelt as much on Japan's economic straits as its environmental problems.

When I arrived in Horiuchi in 1998, Japan was mired in an economic morass (see McCormack 2002), sinking into recession, and was beginning to experience some of the most sweeping social transformations my informants there could remember. "Lifetime employment" (shūshinkoyōseido) has long been a privilege rather than the rule in Japan (see Rohlen 1974; Fowler 1996; Gill 2001; Allison 1994), but even through the weak post-Bubble period there had remained a certain stability in the labor market. About the time I settled in Horiuchi, though, Tokyo was in a deep funk, and a majority of my informants knew of at least someone who had lost a job—a staggering reversal. Few were under the illusion that the windfall conditions of the late 1980s and early 1990s would resurface, but with climbing unemployment, corporate restructuring (risutora), and similar adversities, many despaired over the protracted dreariness of the economic climate.

Major newspapers did little to dispel the gloom. On a typical morning, front pages would offer up a bit more of the bleak pessimism that gripped the nation and its institutions. Such negativism was, indeed, clearly reflected in the grim findings of the influential Tankan survey of business confidence (Bank of Japan 1998, 1999). Bankruptcies, once rare, were becoming frequent both locally and throughout Japan. Nearly everyone I interviewed expressed unease over what the future might bring. While not always in the foreground, the economic difficulties usually lurked in the background of community discourse.

The issues of environment and health loomed large amid this urban unease. Few people believed Tokyo to be a healthy city in which to live, but in the late 1990s, long-simmering misgivings about the insalubrity of car exhaust fumes and other complaints gave way to anxiety over a series of grave threats to health that seized the attention of Tokyo's inhabitants.

Chief among these were waste-related perils, including toxic pollution and environmental illness, which became ready vehicles for community disquiet over metropolitan and national policies as well as indictments of perceived directions in which Japanese society was headed. Before I delve into these important topics, however, I want to describe both the community and certain unusual stresses to which it was subject at this time.

Horiuchi was a more or less middle-class residential community in southern Azuma Ward, Western Tokyo;[6] its calm feel bore numerous similarities to how Izawa residents described their own community before it became riven by conflict over toxic waste there. Horiuchi boasted a beautiful shrine and a prosperous temple, each with lushly forested grounds, but most of the rest of the community consisted of rather densely packed houses and intermittent apartment buildings and condominiums. The area, framed on one side by a massive elevated highway and on two other sides by major thoroughfares, was not far from commercially developed areas with metro stations. However, the ambience of the community remained serene. Children in school uniforms ambled along the narrow, gently curving roads to and from the local schools, neighbors chatted at the side of the road, and locals took walks or exercised along the nearby waterway. Most homes had one or two manicured trees and some shrubbery on their tiny plots, so the area was more verdant than many Tokyo communities, including Izawa. Every evening, even from far off, one could usually hear the sound of the temple gong signaling the end of the day. Horiuchi was therefore not only undeniably urban, part of a teeming megalopolis, but also residential and somewhat village-like due to its sequestration.[7]

One bellwether of Horiuchi's prosperity at the time was the rather anemic traditional-style shopping street that ran through part of the community. The *shōtengai*, as this shopping street was known, was about a quarter-mile long, a quiet road snaking gently from one of the high-volume Tokyo thoroughfares on the outskirts of the community down to the river; its curvilinear route bore testament to its history as a farming track for bringing crops to market and so on—but in terms of commercial vibrancy, this community artery had dwindled to a trickle. Since the *shōtengai* was not contiguous with a major metro station, the strip had been insulated from (or denied) the development that transforms most main shopping streets nearer Tokyo's transport hubs.[8] Furthermore, as a considerable number of the merchants there actually owned their shops and lived above them or nearby, most did not have to struggle to pay high rents. A lack of mercantile frenzy characterized the street, to say the least, and the stagnant economic circumstances, especially during the recessional period of the late 1990s, had a major influence on relations there (see Chapter 7). Shop fronts were narrow and frequently rather shabby-looking, offering Japanese-style sweets, measures of rice in sacks, hand-fashioned tatami mats, fruit and vegetables, electronic goods, traditional medicine, grilled chicken, and so on with a charming hyperspecialization maintained stubbornly in the face of the capital's encroaching supermarket penetration. While zoning regulations had allowed three- or four-story residential buildings to spring up along the shōtengai, the lanes and alleys radiating off the shopping street were tightly packed with houses and the occasional condominium building. Most people walked or rode bicycles to take care of their daily local errands.

On the surface, the tone of sociality on the shopping street was one of polite reserve, interspersed with glimpses of social warmth, industry, and great selflessness. But according to several outspoken merchants and entrepreneurs, the street was rife with frosty exchanges, sniping, and rivalry at this time. Much of the ill will was directed at a handful of conspicuous newcomers with fewer ties of family and shared history than others in the community. Yet it was clear that a sense of fatalism had touched relations in the business community and by extension the tissue of the community itself. Small wonder, then, that underneath the area's veneer of calm, residents expressed anxiety over environmental problems and health issues.

Concerns over pollution and illness hung like a shroud over the tranquil, manicured everyday of Horiuchi, not unlike the haze of urban smog that frequently obscured famous views of Mt. Fuji that Edo and then Tokyo had enjoyed for centuries. On cool, clear winter days, residents of Tokyo can still glimpse the snow-veined peak, a metonym of Japan's natural consciousness and, for some, a pinnacle of Japanese culture (see Berque 1997a).[9] In warmer months, however, air pollution and other climatic haze occlude the view to such an extent that older residents sometimes mentioned the invisibility of the mountain when discussing the ecology of the capital.[10] For the most part, however, residents tended not to dwell overly much on environmental problems per se. These did, nevertheless, color their responses when asked general questions about how they viewed their community, and this gives some insight into links between the environmental and the sociocultural that circulate throughout this volume.

Even before the media furor that serves as the focus of this chapter, Horiuchi reverberated with largely undefined misgivings over toxic pollution and other threats to local health. Residents complained often about the general insalubrity of Tokyo, particularly in comparison to bucolic areas of Japan where they or their relatives might have grown up, bathed as these were in rose-colored nostalgia. It was common knowledge that air pollution led to high incidences of asthma and bronchitis in the capital (see Tokyo Bureau of Environment 2006b), and it was not unusual for someone suffering from a harsh cough to blame the "bad air" in Tokyo rather than, say, their chain-smoking. Local waste incinerators inspired some concern as well. In a society whose postwar history has been littered with pollution debacles (e.g., George 2001; Broadbent 1998; Huddle and Reich 1987; see Chapter 2), and with Japan's conceptions of health informed by East Asian medicine's emphasis on the close interdependent relationship of body and environment (Lock 1980; Ohnuki-Tierney 1984), residents were highly responsive to environmental health stories broadcast in the news. As I moved through the community, I was struck by the myriad ways in which these reports were being retransmitted and often embellished during casual conversation between neighbors, in banter among friends in local bars, in daily admonitions to children, in discussions of housing location and daily hygienic practice, and so on (see J. Thompson 1995; Bird 2003).

Community Gossips

One afternoon in 1998, I interviewed my friend Furuhashi-san, a talkative, insightful woman in her late fifties, next to the small, concrete-banked waterway that snaked through Horiuchi. After about an hour of rambling conversation touching on her childhood in Tokyo and her career as a chef, she began coughing violently, and conversation eventually turned to the insalubrity of Tokyo. "The air is really bad here, terrible. My asthma has been [killing me]..." Then Furuhashi-san revealed an apparent secret: "You know the waste incinerator downriver there? One day, maybe a year ago, smoke came [down this way] with the wind....Everything the smoke touched, the trees and shrubbery around houses here, all turned brown..." Reaching into her handbag for another cigarette, Mrs. Furuhashi then went on, "Some people were all right with it *(heiki)*, but I was thinking, 'This is scary...'" Anthropologists are notorious for shrugging off such scourges as malaria, dysentery, giardiasis, dengue fever, and other dangers they might encounter in their ethnographic peregrinations. Yet I have to admit that I was a little unsettled by the prospect of smoke of unknown toxicity blanketing the community where I had chosen to live. From the small hill where I lived, I could just make out the top of the gigantic incinerator in question over the trees of the local temple—it belched out thick grey smoke more or less around the clock and appeared menacing even from a distance.

For this reason, I made sure to explore this topic a few days later with Miyake-san, a friend of Furuhashi-san's. Miyake-san, owner of a hair salon, made it quite clear that the community was far from healthy. She cited respiratory illnesses she knew about, particularly among children, and also allergies that seemed far worse than when she was a child in Tokyo. But she assured me that she knew nothing about the cloud of incinerator smoke to which her friend had referred. Miyake-san worked in the community, but she lived some distance away, so this qualified her response somewhat. She did in the process, however, comment, almost offhandedly: "Of course, there's Azuma Disease further up north." Azuma Disease. I was stunned to learn, in such a casual manner, that a disease had been named after the very ward in which I had chosen to live. I asked her more about this disease, where exactly it had broken out, how it was contracted, and so on. All she could say was that it was related to air pollution, possibly from waste incinerators. She apologized: "I don't know much about it, actually; ask Mikiko—she knows more [about these things] than I do."

Mikiko wasn't much help on this point, as it turned out, but over the next month my research focused more intently on the threat of toxic pollution in the community. More people spoke to me about Azuma Disease and, since the disease appeared to be named after the ward in which they lived, they wondered aloud whether the problem was endangering their lives and their families, though opinions differed wildly on what the disease was, where it was concentrated, and which areas of the ward were affected. I, too, remained uneasy over the implications of having lived in a contaminated zone for a number of months. My initial knowledge of Azuma Disease in Izawa, then, had been conveyed to me by rumor in very much the same way that many of my informants themselves learned of the affliction. Since media coverage of the Izawa protest was thin, my early inquiries bore a similarly discursive character as I tried to gauge the threat, pinpoint the facility, and gather information on protesters in conversations.

Before long, though, with the breaking of the Tokorozawa dioxin story, toxic pollution surfaced as a daily topic of discussion, propelling toxic pollution to the forefront of many of my informants' thinking. By recording, and at times participating unwittingly in, the same processes of transmission, reception, and retransmission of information that shaped the sociocultural topography of the fieldsite as a whole, I was able to amass important data on human-scale circulation of environmental knowledge.

Many social science disciplines eschew rumor, gossip, hearsay, and informal discourse as unreliable material in favor of "harder" data sources: historical documents, political platform statements, pottery shards, state-generated economic reports, statistical surveys, questionnaires, and so on. In seeking to understand transitory social and cultural milieux, however, fleeting modes of expression are vital—perhaps even more so because their evanescent character leaves them relatively overlooked. I was able to record traceries of discourse both during a period of relative neutrality before the dioxin scare, when informants had less reason to overdramatize their responses, and later, when nearly everyone had some personal investment in finding out about and passing on what they knew of the parameters of danger and damage.

Anatomy of a Media Scare

The conflagration of the Tokorozawa dioxin scare, nearly instantaneous on the rarefied plane of Japanese mass media, did not occur overnight in Horiuchi. Locals who watched the news closely, like Andō-san, were immediately aware of the controversy over contaminated spinach and other vegetables from Saitama Prefecture. But others, such as Sumi-sensei, a kindly octogenarian who gave me free lessons in polite Japanese, were not fully aware of the issue until weeks after the story broke. However, within days of TV-Asahi's initial News Station report, local repercussions of the dioxin scare were palpable. Supermarket chains in the community, such as Keio and Miura, had removed Saitama vegetables from their shelves. And residents, particularly housewives who were around most of the day, participated in an ongoing dialogue with neighbors and friends, local greengrocers, family doctors, and sometimes anyone who would listen as to the extent of toxic risk and how to treat exposure suffered by children, for example. A week after the initial broadcast, when News Station was forced to retract part of their story, the ensuing media storm was mirrored on the local level by retellings, reinterpretations—including some indignant criticism—and other forms of informal discourse.

One reason editors of Japanese news organs seized upon the dioxin story is that they know their audience takes *kōgai,* or environmental damage, very seriously. In this sense, dioxins are similar to such nefarious-seeming, ill-understood vehicles of mediated sensationalism as HIV (Farmer 1992, 1–7, 122) or alleged satanic abuse of children in Britain or America (La Fontaine 1998). Dioxins were "much in the news...complex, not directly observable, and not well understood, even by scientists" (Nickum et al. 2003, 38).[11] Unlike car exhaust fumes, an acknowledged, relatively straightforward threat in Tokyo, dioxins functioned extremely well as a vehicle for environmental fears and misgivings in community Tokyo. In Horiuchi, such toxins were invoked to interpret puzzling events or illnesses, express lifestyle complaints, and vent dissatisfaction with the trajectory of contemporary urban Japanese society.

This interpretive complexity was augmented by the broad reach and variety of mass media involved. Scandalous information inspires transmission through informal means in every social milieu. Yet mass-mediated social knowledge, for obvious reasons, operates at a different scale of magnitude. In addition to a far wider audience—via printing presses, airwaves, copper cables, fiber-optic Internet connections, and "data clouds"—the legitimacy of established institutions involved in transmission tends to give mass media greater credibility. The media scare generated by the Tokorozawa dioxin story tarnished some of that credibility.

Many of my informants found out about contaminated spinach directly from news reports. Yet many more first learned about the prospect of this toxic threat from friends and acquaintances. Naturally, perhaps, the details of the story began to take on a life of their own. Though the flawed TV-Asahi news reports indicated that only vegetables from the city of Tokorozawa were contaminated, soon all vegetables from Saitama Prefecture, and spinach from anywhere, became suspect. The banal produce section at local supermarkets suddenly became enlivened with impromptu debates. When I asked other shoppers about spinach in Keio supermarket during this time, locals warned me that vegetables were "dangerous," "unhealthy," "polluted," and got into debates amongst themselves about the contested details of rival news reports and subsequent rumors regarding dioxins and incineration. Fellow patrons at local restaurants and bars broached the topic with me, worried that I might not know to be careful. Moreover, they changed their habits at home. One housewife in her fifties recalled: "In February...we stopped eating spinach entirely.... We trust [the local grocer] to have safe vegetables and all that. But we just thought that, well, there could be mistakes." Interestingly, one greengrocer on the outskirts of the community was only *initially* careful to make sure that his produce was Saitama-free early on in the dioxin scare. He commented, chuckling, "Everyone was worried about [the dioxins], so I made sure my vegetables came from other areas....But later, when everyone else was still worried about dioxins and the like, I secretly put them back [in my stock] 'cause I knew [Saitama vegetables] were safe." When asked how he knew the vegetables were safe, though, his reasons were less informed than instinctual, allegedly based on his decades of experience with Japanese vegetables. "People get worried about things all the time...I knew the vegetables from Saitama [as a whole] were safe, so why not sell them? They were cheap, too."

Residents sometimes made creative use of dimly remembered news reports or local knowledge to fill in the gaps. For instance, some informants who knew of the environmental protest surrounding Azuma Disease assumed that the culprit there was an incinerator in the center of that community rather than a waste transfer facility. With the upsurge in interest in dioxins, this caused a more pronounced local fear of waste incineration than was probably justified in that area of Tokyo. Others believed that there was actually a protest against the local incinerator some miles downriver from Horiuchi, in Sugibayashi. Despite misgivings among residents there, however, that incinerator was considered state of the art and had a decent reputation for safety.

The attention on dioxin pollution also shone a media spotlight on related health issues that had long lain in something of a grey area of public consciousness. For example, since 1972 the Ministry of Health and Welfare and other institutions had been conducting annual studies on the contamination of breast milk by dioxin and other toxins. These studies had been acknowledged by the Japanese mass media periodically, but public attention lasted only briefly, by and large. Yet with the new currency of dioxins, news organs began to give greater exposure to these troubling breast milk studies and to newer findings, specifying the cities with the worst rates, for instance (e.g., *Yomiuri Shimbun* 3 August 1999; see Ministry of Health and Welfare 1999). Residents of Horiuchi devoted considerable attention to broadcasts reporting that infants can take in up to *twenty-six times* the ministry's "tolerable daily intake" of dioxins merely through consuming breast milk. Some local informants blamed miscarriages and other reproductive concerns on dioxins and other toxic pollution. Of course, toxic pollution is, regrettably, hardly limited to urban Japan, and reproductive misfortunes such as miscarriages and fetal abnormalities befall families all over the world in every conceivable living environment. But the fact that miscarriages and birth defects were being interpreted by some in terms of exposure to dioxins indicates the power of media portrayals—and the resonance of local discourse on invisible threats—in both transmitting and influencing toxic anxieties.

Living the Wastescape

These and other findings suggest that experiences of toxic waste in Tokyo—real and imagined—were shaped considerably by media assertions and local discourse, both of which are prone to distortion, omission, and interpretive license. Accuracy notwithstanding, these greatly influence conceptions of risk in communities. In other words the subjective field of Tokyo's wastescape differs appreciably from a standard geographic representation. In their idiosyncratic mapping of their surroundings, my Tokyo informants were apt to conceive of the city and its environs as a series of zones of greater or lesser risk, whether or not this risk was conclusively proven or, indeed, whether they even understood or could name the alleged dangers involved. Certain hotspots in the region—like Tokorozawa, with its dozens of crude, low-tech incinerators filling Kunugiyama forest and environs with billowing toxic smoke—loomed immensely in the minds of inhabitants due to the toxins they were discharging or were perceived as discharging. This went far beyond longstanding concerns in Japan over living close to a local incinerator. Now, residents of Tokyo knew that even the very food they ate—no matter how wholesome-looking or "fresh"—could be contaminated. Geographical distance and administrative boundaries could no longer guarantee against diffusion of toxins into one's home. At the same time, amid all this concern over smoke, chemicals, spinach, and breast milk, other long-standing public health concerns, such as those regarding automobile exhaust, diminished markedly in significance, whether this change was justified by environmental technoscience or not. The shifting, pulsing, uneven wastescape of Tokyo, discernible in interviews with my Tokyo informants, demonstrated considerable plasticity and was extremely sensitive to new media developments. That first alarmist report on dioxins broadcast by TV-Asahi quickly rippled from one end of Japan to the other, heightening concerns over toxins and waste incineration and buffeting consumption patterns for months.

It would be a mistake to assign too much importance to just one extended media scare. The effects of the TV-Asahi/Tokorozawa dioxin scare were far-reaching and had a substantial impact on public attitudes and even governmental policy documents, but the Tokyo residents I interviewed were also influenced by numerous smaller-scale media-driven episodes of environmental danger and toxic illness. These, taken together, played an important role in shaping denizens' assessment of threats and their decision-making.

During my fieldwork an affliction dubbed "sick house" (*shikku hausu,* or "sick housing syndrome" in English) became a well-known toxic phenomenon in Japan and was prominent in the media. Powerful adhesives and solvents used in building construction contain chemicals that can cause health difficulties after extensive exposure. The problem has worsened in recent years as companies continue to employ new chemical compounds whose full effects remain largely unquantified (see United Nations Environment Programme 2004), particularly in combination. When a family moves into a new house or apartment, chemical residues from construction—if they are profuse enough and if the residence lacks proper ventilation—can seriously impact health.[12] In Japanese discourse, the currency of "sick house," which represented a kind of horror story of domestic life gone awry in the chemical age, meant that Tokyo residents were frequently well aware that toxins could impact their everyday life no matter what their neighborhood or income bracket. Daily exposure to chemicals in Japan was also blamed for such ailments as atopic dermatitis *(atopī)* and asthma *(zensoku)* in children, which seemed to have become far more common in recent years.

With increased consciousness of environmental problems from the mid-1990s on, one health threat that gained wide currency was that of environmental hormones *(kankyō horumon).* Catapulted to notoriety by the environmental polemic *Our Stolen Future* (Colborn, Dumanoski, and Myers 1996)—a critical inversion of the UN-derived report entitled *Our Common Future* (World Commission on Environment and Development 1987), on the project of sustainability—endocrine disruptors that interfere with human and animal reproductive function were identified as being an insidious byproduct of certain forms of plastic products and industrial and petrochemical processes (see also Casper 2003). In Japan, rumors circulated that plastic containers released these hormone-mimicking substances, disrupting human bodily functions, including reproduction, and producing partial feminization of male physiology (see Chapter 7). While many types of plastic vessels were at issue, the most notorious by far were disposable instant-ramen noodle bowls and babies' bottles, since these were either exposed to boiling water or placed in microwave ovens. While dioxins, whose ill effects are surprisingly numerous, are themselves endocrine disruptors, "environmental hormones" are usually discussed by Japanese and reported by the mass media as a category unto itself.

Another notable environmental drama that influenced risk consciousness in Tokyo was the ongoing nuclear mismanagement debacle at Tokaimura. Roughly speaking, Japan ranks second among major nations, behind France, in dependence on nuclear power. This is an unlikely state of affairs, given Japan's bitter experience of nuclear devastation in 1945. Yet despite this traumatic history, Japan not only has an extensive nuclear industry, but it is notoriously accident-prone and poorly managed, and its operations lack transparency. The nuclear energy industry is regulated by the state, but work is frequently contracted out to private corporations like JCO (a now-defunct subsidiary of Sumitomo Metal Mining Company), which had operations in Tokaimura, only about eighty miles northeast of Tokyo. On 30 September 1999, residents of Tokyo learned that a "criticality"[13] incident had occurred at JCO's Tokaimura facility. Work at the facility was cloaked in secrecy due to security issues related to uranium processing. Nevertheless, it soon became clear that poorly trained workers had opted to save time by skipping key safety steps and had dumped a bucket of liquid uranium into a settling basin in a quantity that greatly exceeded safety limits. The ensuing chain reaction represented Japan's worst peacetime nuclear debacle and the world's most serious nuclear accident since Chernobyl (JCO Criticality Accident Comprehensive Assessment Committee 2002; Kerr 2001; see also Petryna 2002).[14] The accident and resulting fire were eventually brought under control, but three workers were critically injured in the process (two later died) and forty-six other employees were exposed to serious levels of radiation. Throughout, Tokaimura residents were ordered confined to their homes, with doors and windows shut to the contaminated air, but more than a hundred people in the community were found to have been exposed to low-level radiation (Aldrich 2008; Kerr 2001; JCO Criticality Accident Comprehensive Assessment Committee 2002).

As the incident unfolded, residents of Tokyo—living not far beyond what, in Japan, is deemed reasonable commuting distance from this potential disaster-in-progress—were gripped by the news and wondered whether they could afford to believe the government's assurances that the problem was contained. I myself toyed with the idea of taking the bullet train out of harm's way to Osaka.[15] Nuclear radiation is relatively easy to detect, and it was soon clear that Tokyo's population was spared a nuclear calamity. But the seriousness of this accident, along with the facility's proximity to Japan's most populous conurbation, put Tokaimura on the map, so to speak, in Tokyo's risk consciousness.

These episodes suggest that toxic pollution as *lived* is not just a simple sequence of cause and effect—environmental poison and subsequent illness—but a complex, subjective phenomenon that must be analyzed qualitatively. When we consider how threats are experienced, human spatial reckonings do not conform much (or at all) with a standard map. People's memories, experiences, and navigations, along with mass media, hand-held communication devices, and informal discourse both inside and outside a given community, cohere into a complex discursive meshwork that influences (and recursively responds to) the beliefs and fears of denizens as they conceive of and engage with their surroundings. For Japanese, this commingling of cognitive and concrete landscapes, this anxiety over imagined toxins on the one hand and actual defilement and devastation on the other—what I term Japan's wastescape—constituted part of an ongoing process of negotiation informed by history, language, and ontology. Dioxins and other environmental risks were "in the air," so to speak, pondered and debated and feared in communities. Japanese engagement with "environments" included visceral responses to toxic waste and environmental illness, attitudes toward "nature," hygienic practices and social exclusion, trust in (or suspicion of) the state, reproductive anxieties, and perspectives on waste, excess, and sustainability that were exceedingly Japanese in character. From the "Industrial Waste Ginza" of Tokorozawa to the relative safety of my Horiuchi fieldsite, Japanese interpreted their surroundings in idiosyncratic ways that nevertheless remain important to understanding both toxic environmental problems and Japanese social life generally.

CHAPTER 4

The Cult(ures) of Japanese Nature

> On such a springlike day as this,
> When the light suffuses
> Soft tranquility,
> Why should the cherry petals flutter
> To the earth so restlessly?
> —Ki no Tomonori, *Kokinshū 84*

Chapter 3 focused on attitudes toward environmental risk among my informants in Horiuchi. Though this portrait of everyday environmental attitudes and behavior provides a contrast to the somewhat more charged political space of my Izawa fieldsite, it furnished only one, albeit important, ethnographic dimension of life in community Tokyo. For much of what shapes environmental engagement in Japan derives from conceptions of nature and interventions into natural-cultural landscapes there.

In Japan codified nature-focused activities and eco-symbols—manicured gardens, *ikebana* flower arrangement, paper houses, and so forth—give a veneer of apparent ecological sensitivity to Japanese social life. Occasionally they flag a sincere and abiding concern with fleeting shifts in weather, surroundings, and wildlife. But more often they served to deflect attention away from concrete lived surroundings and ecology and contribute to a pervasive form of detachment in Japanese urban communities. Japanese love of nature, as well as the Japanese *idea* of Japanese love of nature, revolves partly around nostalgia for what are considered traditional relations with nature (that is, village life in an agrarian community). But this wistful, retrospective construction of Japaneseness also manifests itself in occasional local festivals and community events that mark the seasonal cycle and serve as rites of commemoration of community identity. Perhaps the most notable of these is the annual frenzy surrounding cherry blossoms *(sakura)*.

Hanami (literally, "flower-viewing") seizes the nation every spring. The focus of attention, particularly media attention, is the brief efflorescence of cherry trees. *Mankai,* or full bloom, lasts only a short time. And due to idiosyncrasies of climate, contiguous areas may at times find themselves at different stages in the strictly coded progression of the *sakura.* Television news programs chart the *sakura's* progress across the nation with color-coded maps. Print media publish photographs of famous, centuries-old trees or copses of flowering cherry trees in striking landscapes, and advertisements featuring *sakura* bedeck trains, billboards, gigantic building-side television screens, and other sites of hypersemiosis. Early in my fieldwork, the national bank Sakura (strategically named and now defunct), stepped up its commoditization of this natural-cultural symbol with huge, floral advertisements at a branch not far from Horiuchi. In public parks and on concrete riverbanks all over Japan, groups of friends, colleagues, and relatives descend on stands of cherry trees in order to drink and eat at night under the radiant, illuminated blossoms. In Tokyo parks well known for older stands and well-composed vistas of *sakura,* nearly every patch of available ground might be claimed by seated groups of revelers seated in stocking feet on tarpaulins, and it is frequently difficult to walk except on the crowded paths.

Each year, during my fieldwork, Horiuchi marked the coming of the *sakura* with a flower-viewing barbecue in a grove of cherry trees near the river. Residents counted themselves lucky to have this resource at their disposal; the stand of five mature *sakura* was located on land owned by the nearby Buddhist temple, and it was through the temple's largesse that use of the land was granted to the *chōkai,* or community association, for the morning and afternoon. A high chain-link fence ringed the grove to prevent interlopers from gaining access to this rare Elysium. Cherry trees also lined the river, their snowy boughs drooping over the concrete-lined channel, and Japanese with cameras would stroll down the riverside paths, often posing for photos against the picturesque backdrop. *Chōkai* members like Fukushima-san spent the evening and morning preparing for the approximately two hundred residents or so who came to spread their blue-plastic tarpaulins and blankets under the cherry trees. Those residents who turned up consisted almost exclusively of families with children, elderly groups, and the organizers; teenagers and unmarried young professionals tended to prefer sites like the trendier, more scenic and romantic Inogashira Park, also located in Western Tokyo.

In contrast to more boisterous celebrations around Tokyo, Horiuchi's *hanami* was a muted affair. Adults drank beer, *shōchū*, and *sake*, but the emphasis was on relaxing in small groups and chatting over lunch.[1] Like children the world over, the kids wandered through the snow-white blossoms on low-lying branches, playing around them and sometimes nearly destroying them. More than a few residents had their cameras out to commemorate the occasion, taking pains to arrange compositions of friends/relatives against downy backdrops and sometimes trying to edit out nonnatural features, like a pickup truck used by the organizers. However, most were engaged in *hana yori dango*. The phrase, used with self-irony, captures the often clear priority in Japan of social relations in nature over the nature itself. Meaning literally "more the dumplings than the flowers," *hana yori dango* evokes how Japanese often prioritize the social exchanges taking place under the trees (metonymically encapsulated in grilled *dango* dumplings, common at *hanami* events) over the blossoms that comprise the putative focus of the gathering.[2] Indeed, Japanese engagement with nature often resembles a person holding aloft two mirrors: one with which to regard the nature in question, and one for looking at *oneself* looking at nature.[3]

Sitting under the trees, however, did immerse residents in one much-lauded facet of the *sakura*'s beauty: the falling petals. Cherry blossoms are praised in Japanese cultural production for their poetic transience. By the time a grove of trees like Horiuchi's has reached full bloom, even a light gust of wind will send white petals cascading down like snowfall. The shedding is gradual but continuous, and after a few hours, tarpaulins, clothes, and other belongings can be carpeted in the white petals. Though at times overblown, this prized ephemerality is central to Japanese appreciation of nature and accounts for the privileged place *sakura* hold in Japanese cultural production.[4] Falling cherry blossoms recall, for Japanese, the impermanence and fragility of life, which is resonant in a society with a long tradition of adherence to certain Buddhist principles (see LaFleur 1992), as well as such related themes as the tragic death of the young, virginity (and its loss), and, notoriously, the fate of kamikaze suicide bombers in World War II (e.g., Ohnuki-Tierney 2002; Hoffman 1986).[5]

In understanding a society's environmental engagement, that society's views on nature loom with significance. This chapter explores the considerable extent to which Japanese lavish discourse on certain highly standardized forms of nature while preferring to ignore, or at least obscure, elements of their own ecology such as waste and pollution, or irksome animal interlopers like vermin jungle crows, that contradict the rhetoric.

"Nature" in Japanese

Raymond Williams calls nature "perhaps the most complex word in the [English] language," and he identifies three core meanings, which are sometimes in direct contradiction with each other: "(i) the essential quality and character *of* something; (ii) the inherent force which directs either the world or humans beings or both; (iii) the material world itself, taken as including or not including human beings" (Williams 1988, 219; original emphasis). As the definitions make clear, there are numerous possible English sentences where conflicting meanings intermingle without dissonance to native ears. Due to its widespread usage and its diffusion into politicized terms such as "innate," "native," and "nation," understanding nature is key to decrypting a society's attitudes toward surroundings, identity, intermarriage, immigration, and so on.

Given the socioculturally anchored associations of "nature" and linked terms in English and other related languages, it is not surprising that Japanese presents further variations on a theme. The most common Japanese translation for the English word "nature" is *shizen*, originally adopted from the Chinese approximately 1,500 years ago (Tellenbach and Kimura 1989, 153–155; Kalland and Asquith 1997, 8). With reference to Williams' three core definitions of the word "nature," *shizen* covers only perhaps the latter two: nature as inherent force or order, and nature as signifier for the material world. (Predictably, neither does the English word capture all the meanings of the Japanese term.) In addition, the Japanese character *sei* tends to correspond with Williams' first definition: nature as the "essential quality and character" of something. To *shizen* and the character *sei*, we can also add *tennen*, another Chinese loanword denoting more concrete and material ecological elements (as in *tennen shigen*, or "natural resources," which strays closer to Williams' third definition), though it was used less than *shizen* by my informants.

Mizukara and *onozukara* provide intriguing examples, since these ancient words (still lingering somewhat in modern Japanese usage) together apparently used to connote "nature" before the word *shizen* was adopted from the Chinese. Some scholars of Japanese nature (Kalland and Asquith 1997; to some extent Kyburz 1997; and Berque 1997a regarding *onozukara*) believe that *mizukara* and *onozukara* provide a glimpse into the autochthonous natural-cultural psyche of the Japanese. Kalland and Asquith (1997, 10), in their introduction, define *mizukara* as " 'oneself' " and *onozukara* as " 'what-is-so-of-itself,' " and this dual semantic usage reflects, for them, a complex conceptualization of nature that blurs the distinction between individual and environment in Japan. These shades of meaning, though attenuated, still resonate with some present-day connotations: *mizukara ni,* in the adverbial form, may index an act that an individual engages in self-consciously and intently, something springing from the will of the person.[6] *Onozukara ni,* which shares a root with *mizukara ni,* connotes a spontaneous, irrevocable natural act, such as the changing colors of autumnal foliage, over which humans typically have no control. I find such meditations on these two ancient words revealing, but also overwrought. Such anthropological analyses are also uncharacteristically speculative, partly since *mizukara* and *onozukara* surface rather rarely in contemporary usage.

Still, "nature" (variously defined) pervades contemporary Japanese society. Along with the pervasive notion that Japan is a homogeneous society with distinctive universal characteristics, such as groupism, and a uniqueness that distinguishes Japan from all other nations, Japan resonates with the belief that the singular characteristics of Japanese culture derive from the particularities of Japanese nature (Berque 1997a, 105; Dale 1986). This rendering of "Japanese nature" has thrived within the quasi-scholarly genre of conservative, essentialist Japanese writing known as *nihonjinron* (or "discourses on the Japanese") and has influenced many in a country prone to collective introspection. A passage from Peter Dale's trenchant *The Myth of Japanese Uniqueness* illustrates the vast scope of this discourse:

> Outsiders may have difficulty coming to grips with the precise dimensions of the nihonjinron [sic]. Just imagine the situation which might ensue had English letters over the past 100 years been singularly preoccupied with the clarification of 'Englishness', not only as an essayistic form but as a major subject of austere academic research. Imagine then dozens if not hundreds of works pouring from the presses of Oxford and Cambridge...treating everything under the English sun as consequences of some peculiar mentality unchanged since one's ancestors first donned woad and did battle with Caesar; imagine this as something which filtered down through newspapers and regional media to everyday life, and you have something of the

picture of what has taken place in Japan, where almost any discussion
from the formally academic to the colloquial market-place exchange
can reflect this ideology of nationhood. (Dale 1986, xii)

The tone of Dale's comments suggests some exaggeration and may seem
even flippant, but what is striking about postwar Japanese discourse on
Japanese identity is that Dale's is a relatively evenhanded assessment.[7]
Naturally, Japan is not the only place where national-cultural imaginings
have flowered (e.g., Pevsner [1956, 1968] on "Englishness"), though it is
clearly more respectable for Japanese to dwell on them.

Ideologically blindered *nihonjinron* writers discern the Japanese
people's stereotypical traits as stemming from immersion in the sequestered Japanese natural milieu. For example, as the *nihonjinron* argument
often goes, a history of benevolent (though changeable) climate and
an abundance of natural resources predisposes the Japanese to lead a
peaceful, nonaggressive, collective agrarian existence in harmony with
nature in contrast to carnivorous foreign peoples who slaughter live-
stock, fight internecine wars, and grapple with a harsh environment
(Tsukuba 1969; Berque 1997a; Dale 1986, 41–46).[8] This "uniqueness"
of Japanese nature also resides in the Japanese body: in the face of evidence that Japanese and Koreans share ancestral roots and that even the
Imperial House bears some foreign heredity (e.g., Dale 1986, 43; Diamond 2005), essentialist Japanese scholars argue that the Japanese "race"
extends back through the millennia in an unadulterated line. This difference is usually characterized in terms of blood. *Nihonjin no chi,* or
"Japanese blood," signifies the inherent Japaneseness of natives and the
undeniable otherness of foreigners (Yoshino 1992, 118, 22–32).[9] Thus,
by extension, the Japanese gene pool (an aspect of Japanese reproductive nature), sheltered in this island country *(shimaguni),* is positioned
as utterly unique and pure when compared with the exogamous "alien
'horseriding nomads'" and the like who allegedly comprise the blood
mix of non-Japanese (Dale 1986, 43).

The work of Japanese philosopher Watsuji Tetsurō (1889–1960)
lies at the center of this discourse, for his influential concept of *fūdo*—
translated as "climate and culture" or "milieux" (Befu 1997, 106; and
Berque 1997a, 40, respectively)—is emblematic of such notions of the
ecological foundations of Japaneseness.[10] Watsuji's *fūdo* indexes "the
natural environment of a given land, its climate, its weather, the geologi-
cal and productive nature of the soil, its topographic and scenic features"
as well as its culture and history (Watsuji 1962, 7; Watsuji 1961, 1).

Prominent Japanese scholars before Watsuji had examined Japan's fixation on nature, arguing that ideas about Japanese culture were inspired by Japan's natural beauty (Shiga 1937 [1894]) or by Japanese people's love of nature, as in Haga Yaichi's 1907 "Ten arguments on [Japanese] national character" (Minami 1994; Befu 1997, 106–107). With a dash of philosophical rigor, Watsuji surpasses these flawed accounts somewhat in positing that human existence in a "climate" has both spatial and temporal dimensions and is inseparable from the subjectivities of the human body (Befu 1997, 110–111; Berque 1997a, 40–45). In grappling, often clumsily, with "climatic" influences on Japaneseness, Watsuji's work represents a domestically influential Japanese account of engagement with surroundings. Watsuji wrote *Fūdo* as a response to Heidegger's *Being and Time* (1927), and one of his key criticisms of Heidegger is that the German frames human existence as an individual condition, without due consideration of the relationship that an individual has to others in a social milieu. Watsuji, from his holistic Japanese anchor-point, believes that the human condition is one of "betweenness" *(aidagara)*. To grasp this, one need only recall contemporary Japanese invocations of the sociality of cherry-blossom viewing. Persons live life bound together in social networks, and collective experience in a "climate" drives Watsuji's emphasis on *fūdo;* while individual time ends with death, collective social existence embedded in customs and mores—and influenced by climate—lives on (Yuasa 1987, 170). "Betweenness" is shaped by spatio-social topographies, and for Watsuji, engagement in environs must be relational; indeed, the milieu has the effect of "environ-ing" social relations that transpire there.

The wide dispersion in Japan of Watsuji's ideas regarding nature and climate make his theories of value to understanding Japanese engagement.[11] His work is, of course, tainted by the same problems as those of his intellectual *nihonjinron* kin. Yet despite their circular logic and two-dimensional portrayal of the Western (not to mention the Japanese), ideas about Japanese nature (and Japanese natures) surface frequently in newspaper editorials, letters to the editor, talk shows, blogs, and the like. Perhaps equally often, such politically charged ideas of Japaneseness came out in discussions with members of the communities where I conducted research, their thinking and behavior having emerged from the ideological crucible of Japanese educational institutions, media, and discourse. More than a few of the residents of Horiuchi I interviewed brought to my attention, in almost conspiratorial tones, the fundamental difference between Europe, a hunting culture *(shuryō),* and Japan, a farming culture *(nōkō),* thereby hinting at the vast gulf between our worlds. I cannot count the number of times that a Japanese person has indicated that Japan is unique in the world due to its having four distinct seasons. The four seasons *(shiki)* concept in Japan, though clearly erroneous when compared my hometown in Vermont, serves to emphasize the singularity of the Japanese climate (a popular *nihonjinron* theme) and to explain the minute attention many Japanese pay (or imagine they pay) to the vicissitudes of the weather and the turn of the seasons.

In this way, environmental engagement in urban Japan cannot, I believe, be properly understood save in the context of Japanese conceptions of nature, interventions in "natural" spaces in urban confines, discomfort with waste-scavenging vermin (see Chapter 5), anxieties over pollution (see Chapters 6 and 7), and competing views on the path that Japanese society has taken.

Consuming Nature

It goes without saying that the "climate" of Japan has departed a great deal from the utopian *fūdo* that Watsuji and other describe in their writing. Indeed, in hyperdeveloped, media-deluged cities like Tokyo, Japanese fascination with "Japanese nature" has become codified to such an extent that even enthusiastic social interventions into Japanese natural milieux can carry an element of detachment typical of contemporary urban Japan.

Japanese discourse on nature partly involves nostalgia for what are considered traditional relations with nature—that is, village life in an idealized agrarian community. With more than three-quarters of Japan's population now living in urban areas, pining for the lost poetry of village life has taken on vast dimensions and is interrelated with urban community-building initiatives (Robertson 1991; Bestor 1989). A frequent vehicle of this nostalgia is *furusato,* whose characters index "old village," but which might better be translated as "native place" (Robertson 1998, 110).[12] *Furusato* is a protean phenomenon and therefore has different registers of significance for Japanese, yet this is, by and large, rural nostalgia in the abstract. Rather than longing for one's actual hometown or birthplace, which is very likely not a bucolic, idyllic hamlet at all but perhaps a dense collection of apartment blocks in a stretch of urban sprawl, "the word is used most often in an affective capacity to signify...rather the generalized nature of such a place and the nostalgic feelings aroused by its mention....a warm, fuzzy, familial, and ultimately maternal aura" (Robertson 1998, 115; Robertson 1991; see also Ivy 1995, 103 passim).

Many of my informants had particular *furusato* outside Tokyo that they visited annually, usually where they or their parents or grandparents had been born. Many other informants had lived in Tokyo for decades and considered themselves transplants from elsewhere for career reasons, explaining that they were originally from the mountains on the island of Shikoku, say, or from picturesque Nīgata Prefecture on the Sea of Japan. Nevertheless, most of those with "real" *furusato* also participated in the more generalized longing that Robertson describes.

While assuagement of this diffuse yearning often leads to domestic tourism and commoditization of the village as a crucible of nature and tradition (see esp. Ivy 1995; see also Moon 1997), many Japanese get their fix by participating in local festivals or visiting local bars and restaurants that ooze with rural nostalgia.

SERENADING THE VILLAGE

Some weeks after I moved to Horiuchi, an acquaintance who belonged to the community association urged me to sign up for the karaoke contest scheduled the following weekend at the annual harvest festival *(shūkaku [no] matsuri)* in September 1998. I already knew that Japanese take karaoke very seriously, and I politely demurred. Knowing I was in the market for a bicycle, though—I was, at the time, riding a small, pink "granny bike" *(mamachari)* loaned by an informant's teenage daughter—he continued to twist my arm: "C'mon, if you sing a Beatles song, you'll probably win the mountain bike! Besides, it's cheaper if you sign up through me." Eventually I relented. But I wanted to sing in Japanese, and I decided to consult Mama-san, an early supporter in the community. Knowing something of my research interest in Japanese invocations of nature, she exclaimed, "I've got it—I know just the one for you. 'Kitaguni no Haru' (Spring in the North Country). Absolutely *(sō sō sō)*. It captures that mood perfectly." The next day, she gave me three CDs of traditional ballads *(enka)*, one containing her recommendation, As I began memorizing the lyrics, I could not help but notice that it was, indeed, a good, even witty, choice. An excerpt from the song gives an impression of the genre:

> White birches, blue sky, a southerly wind. Trees blooming [on] a distant hill—ah, spring in the north country *(kitaguni)*. In the big city, you don't really understand the seasons.... What's that girl [from my hometown] doing now? I wonder if we will ever go back to that village *(furusato)*. Perhaps we should go back.[13]

The song encapsulates many of the nostalgic elements in other songs performed in the contest (and sung in karaoke bars all over Tokyo)—note in particular the appearance of the term *furusato*. Though some songs are less overtly rural nostalgic, many exude a sense of loss and a longing for a Japan that most feel is a thing of the past.

Some of this pathos was voiced during the coming festivities. The harvest festival spanned a full weekend and took place on shrine grounds, just a stone's throw away from the grove of cherry trees. The grounds cover more than an acre on the side of a hill leading up from the waterway's edge, at the top of which sat the ornately constructed and carved wooden shrine named Kaeru-no-jinja. The succession of stairways leading to the front of the shrine was demarcated by three large, *pi*-shaped gateways *(tori)* made of either granite or wood. Also punctuating the route were a number of large stone lanterns and two pairs of stone statues depicting mythical dog-like guardian beasts *(komainu)*, each pair straddling the path. Though some open space existed in front of the shrine buildings, the overwhelming remainder of the compound was covered with towering trees, and their thick canopy kept the grounds in dappled shade.[14]

The festival activities were rather clearly bifurcated between daytime and nighttime. Early on Saturday, local families came to bring ritual offerings of *sake* and other rice products to give thanks to the *kami* (spirits).[15] Stalls selling *matsuri* foods—ice cream, cotton candy, and so on—commingled with stalls offering games for children—one being a goldfish hunt.[16] During the day, traditional Nō drama was performed on a stage in a wooden shrine building. Later, a group of men wearing *happi*-robes (and fewer women) took part in the *mikoshi* portable shrine procession. Though the entrance of the *mikoshi* into the precincts of the shrine was an affair of great spectacle and ceremony, it was overshadowed by the events of the evening.

Separate karaoke contests for Saturday and Sunday nights served as the festival's central entertainment. As is typical with karaoke, the musical entries were dominated by *enka*, a commercialized genre of love songs and dirges distinguished by syrupy, rural-nostalgic lyrics that hark back, for example, to the "north country" *(kitaguni)*, to the mountains, to village life, and to *kokoro*, or "heart," a quality sometimes perceived by my informants to be at some scarcity outside the bonhomie-filled confines of the leisure space. During the night's competition, which about forty singers entered, hundreds of spectators watched the singing and ate, drank, and socialized under the trees amidst Shinto artifacts—a fleeting re-creation of the village of yore in the heart of one of the community's most sacred (and usually rather desolate) spaces. One after another, the contestants, who spanned a wide range of ages and occupations, serenaded this mythical village with soulful ballads invoking tragic loss, broken hearts, changing values, and narratives of fickle destiny.

When I reached the stage, the audience was clearly surprised and amused at the spectacle of a foreigner warbling nostalgically about the hinterland in Japanese. By the second verse, the entire congregation of about three hundred "villagers" joined in, singing the remaining two verses with me and clapping out the rhythm. First of all, this demonstrated some familiarity with the song. And the trial-by-fire seemed to serve the purpose of incorporating me symbolically into the community, at least superficially. Numerous times in the weeks following the contest and throughout my fieldwork, residents approached me, harking back to my performance. This goodwill, alas, did not extend to awarding me the mountain bike, but I did win a juicer.

RAISING THE RED LANTERN

Annual events such as *hanami* and the harvest festival are emblematic of Japanese natural-cultural predilections and the character of Japanese engagement with surroundings, but they arrive only intermittently during the cycle of the seasons. More commonly, my Japanese informants celebrated qualities of "Japanese nature" and rural village life in a range of thematized leisure domains that reproduced and interpreted these fetishized eco-symbols in settings that cater to Japanese social practices.

During my initial fifteen months of continuous fieldwork, as during numerous shorter research periods, my structured and semistructured ethnographic interviews took place in private circumstances and with plenty of time for my informants' embedded social knowledge to emerge. "Participant observation," on the other hand, frequently took place at times of leisure, when people in the community, free from the pressures of work, congregated and participated in activities and discussions in which they felt they could relax. In addition to meals at private homes, tourist excursions, and visits to theme parks and an artificial indoor beach domain, I also joined informants in a very popular traditional-style bar located on the main shopping artery in the community.

Chikurin, as the bar was called, was an ideal ethnographic arena: it boasted a loyal clientele that spanned socioeconomic categories and served as an informal meeting place for locals who sometimes had daily obligations all over the Tokyo area and beyond. This *akachōchin,* or "red lantern" restaurant,[17] had tiny dimensions but made up for lack of space with camaraderie. Chikurin could seat a maximum of about twelve patrons around an L-shaped counter, behind which the *mama-san* (proprietress) cooked, poured, served, cracked jokes, stimulated conversation, lent a sympathetic ear, and generally created a warm atmosphere for her visitors. "Mama-san," as the proprietress was known, was a shrewd, highly charismatic woman who had been a stage actress in Kobe, as well as in Paris for a time; she immediately understood what I was trying to do anthropologically and became immensely helpful in facilitating contacts and interviews, explaining Japanese customs and local complexities, and so on. Mama-san described the appeal of her establishment in these terms: "Everyone spends the day working; they're very busy all the time....[Chikurin] is about communication *(komyunikeshon).* Connection *(tsunagari)....*Here, I think we've succeeded in that sense, everyone feels connected. They can tell stories, talk about the day, and then go home and go to work the next day." Mama-san's job, then, was to create an environment where her patrons could feel at home away from home—an environment qualitatively different than those elsewhere in the megalopolis in which they spent the rest of their day.[18]

Nature here was key. In accordance with the nostalgic, bucolic associations of the *furusato* metaphor, Chikurin was by far the most lushly adorned storefront along the commercial artery. (Significantly, *chikurin* means "bamboo forest.") Black bamboo, ferns, and other verdant metonyms for the natural rural village enwreathed the bar/restaurant in a display of thriving greenery amid depressed local surroundings. This cultivation of the natural continued when the patron crossed the threshold. Inside, amateur *ikebana* flower arrangements decorated each end of the counter, often arching upward nearly to the ceiling. The flower arrangements changed in tandem with the progress of the seasons. In March, for example, Mama-san displayed a few hothouse-cultivated branches of *ume* plum blossoms, several weeks before these trees bloomed outdoors. Soon after, a bouquet of *sakura* cherry blossom boughs drew lavish praise from patrons not long before the cherry trees by the river burst into their canopy of white. Though some of Mama-san's floral arrangements were chosen for their visual aesthetics alone (and, being an artistically inclined woman, Mama-san was averse to restricting herself to any set of rigid design rules), most had some direct relation to the change of the seasons and played off of traditional conventions surrounding Japanese expressions of the natural.

The décor of numerous bars and restaurants all over Japan betrays the evocative power of nature, tradition, and nostalgia. While there is no need to catalogue the many examples of folk crafts *(mingei)* and rustic wall hangings inside, suffice it to say that Chikurin was a veritable shrine to "old village" sensibilities. When patrons passed under the necklace of illuminated red lanterns hanging outside (inscribed with the names of loyal regulars), they received a warm welcome and found themselves enveloped by visual and olfactory cues that they had stepped into a different realm. Naturally, mawkish *enka* ballads (or sometimes jazz) were piped in on a cable music system as a finishing touch.

Such establishments are frequently as much about social interaction as about alcohol and food. A big draw for patrons of Chikurin was Mama-san's personal touch. Mama-san's skill was in making Chikurin a haven of community camaraderie and cheer, an alternative Japanese community that helped dispel the pettiness that darkened some areas of Horiuchi's checkered social topography at that time (see Chapter 7). Mama-san was adept at keeping humorous conversation flowing, sometimes defusing a heated debate, more often getting acquaintances to engage in pleasantries or general discussions with her or others on a topic of general interest. Most guests described the warmth of the bar in terms of *kokoro,* or "heart"—one used the English word "family"—and a core group of regulars credited Chikurin (and Mama-san) with being the catalyst four years earlier for their tight network of friendships across what are usually socioeconomic gulfs.[19] In a social milieu reverberating with rhetoric over idealized village camaraderie, Chikurin was an example of real community relations in the city.

Another, somewhat subtler, commentary on the natural is the seasonal cycle of dishes offered in many Japanese-style *(washoku)* restaurants. The Japanese, on the whole, take their food very seriously and prize dishes with ingredients that are *shun,* or "in season." Gourmands and more subdued Epicurean Japanese alike, usually influenced by television cooking shows and glossy magazine spreads heralding traditional Japanese ingredients, are well aware of the annual cycle of which vegetables, fruit, and types of seafood come into season (cf. Moeran 1995, 123–125). Cuisine featuring ingredients "at the peak of freshness" was at the center of Mama-san's fare. Customary small dishes that serve as appetizers when patrons first sit down in an *akachōchin* may consist of a modest pile of marinated and seasoned *shun* green beans, or a few small pieces of a fish that has just become *shun.* At more upscale restaurants, great priority is placed on *hashiri* ingredients—that is, those just in season, perhaps the first taste of a certain delicacy that a customer would have had that year.[20] In discussions over the four seasons *(shiki)* concept, informants like Fukushima-san sometimes rhapsodized about the importance of the seasons as reflected in Japanese cuisine: "As the seasons progress, so does Japanese cooking. Ingredients come in season, and the food we cook changes with it.... That's why we look forward autumn, since there are so many good-tasting foods in autumn." Since Japanese can literally taste the seasonal cycle, so this reasoning goes, they pay closer attention to nature than foreigners do. Though there may be a small grain of truth to this, and while many urban Japanese do appear to attend closely to fluctuations in climate, it would be difficult to conclude that they pay closer attention to the weather than, say, the English do.[21]

Suffice it to say that Japanese do appear to lavish intensive discourse on certain highly standardized and mediated forms of nature. My interviews with informants revealed that while many of them could name foods that were in season at the time the questions were asked, only some could name more than a handful (if any) that came next in the cycle. Cooking shows and magazine stories, as well as the fare down at the local *akachōchin* and at exhibition sites like food courts in department stores, seemed to play a key role in whipping up excitement for these prized ingredients. Though the prevalent appetite for *shun* and *hashiri* foods is not just a commercial invention, there does appear to be a supply-side manipulation of the trope of "Japanese nature" to boost consumption of the material and the symbolic.

This widespread Japanese fixation on cuisine, and gastronomic consumption of Japaneseness, gives one indication of why the Tokorozawa dioxin scare had such broad repercussions. In early February spinach is not really *shun*, but when supermarket chains dropped Saitama produce from their shelves, restaurants and bars like Chikurin reacted swiftly, some choosing not to prepare spinach for fear of alienating worried patrons. Mama-san described her feelings on the matter like this: "It's terrible, this *kōgai* (environmental damage).... Next year I'll offer spinach, and guests will eat spinach. What I wonder is, will it taste as good?" In this way, the unnerving side effects of toxic pollution (imagined or not) as well as the poignancy of associations with the natural milieu make the topic of food and its contamination a more politically sensitive topic than it might initially seem to a non-Japanese. Indeed, while the report by News Station centered essentially on spinach from Tokorozawa, the feared contamination of vegetables by dioxin pollution would not by any means be limited to spinach. It is revealing that my informants tended to speak of vegetables *(yasai)* broadly, rather than just spinach *(hōrensō)*, when discussing their feelings about such toxic pollution.

With both the festivals described above and with other Japanese natural-cultural interventions, there remained a certain sense of the spiritual in encounters with nature, however attenuated during leisure activities. Contemporary urban Japan is a mostly "disenchanted" milieu where the animism of Shinto, for example, plays a less prominent role in people's lives than it did in the lives of their forebears. Yet appreciation of nature verges on worship of nature in many native articulations and practice, even though group conduct in natural milieux usually belied this reverence. In this sense Japaneseness resembled a kind of secular religion—for this was "Japanese nature," and residents were displaying their Japaneseness in participating in these and other activities. As they held aloft the two mirrors referred to earlier—one for regarding Japanese nature, the other for appreciating their Japaneseness in doing so—they participated in an engagement that implied some detachment, as from an awesome entity. Revealingly, in an interview discussing relations with nature, one informant who was a part-time sculptor, Moritake-san, expressed some surprise when I asked whether he ever spent time in shrine or temple grounds. I had mentioned (disingenuously) that I some-times did stretching within the sylvan shrine grounds and occasionally sat there to relax. He responded, "Really? . . . For Japanese, I think, [those woods] have a deeper significance." For him, it was these woods' spirituality that kept Japanese at a remove. Taking this detached reverence for some forms of nature into account, we can see, then, that much of the remaining greenery in residential urban Japan comprises such "sacred woods," with their tall trees embowering ritual spaces. While huge tracts of rice paddies and other secular land have been converted into housing or commercial developments in communities like Horiuchi—creating many a local fortune and adding some *narikin* (nouveaux-riches) to the community—shrine and temple grounds were usually preserved and remained oases of verdure. It seems, then, that despite the many sacrifices made for industrial growth in postwar Japan, particularly environmental sacrifices, some things were "still sacred."

The Power of the Fleeting

Examining the different points of intersection between conceptions of nature and rural nostalgia, perhaps the most telling commonality is the consistent prizing of the ephemeral in Japan. Whether it is the nationwide drama that surrounds the brief efflorescence of cherry blossoms each spring, or the narrow temporal window in which certain culinary delicacies are deemed "in season," or the always-vanishing, but never quite vanished (see Ivy 1995) quality of that fundamental, "unique" natural-cultural village experience, many Japanese fixate on the fleeting when remarking their natural milieu. If there is something of a religious quality to "Japanese nature," then this spirituality is enhanced by its ephemerality. There does seem to be a feedback-loop operating here, for the very transience of the nature prized by Japanese makes it even more prized. The rarer it is, the more special it is. And the more prized transient nature becomes, the more aggressively it is consumed, with some of this mass consumption marring the natural-cultural sites in question. This, coupled with a Japanese predilection for certain highly culturally mediated natural forms (cherry blossoms, autumn foliage, even Mt. Fuji from afar), portrayed constantly through representation in various media, fashions a milieu where nature is still prized in the absence of its conspicuous referent. Though not all-encompassing by any means, this social disposition towards nature does help explain why, in a society so besotted with the idea of harmony with nature, much environmental defilement of the natural milieu goes uncontested except in local contexts. For my urban Japanese informants, close identification with mediated natural-cultural forms seemed to diminish identification with specific "wild" sites outside their homes and *furusato* that development or pollution could corrupt or sometimes already had corrupted. While members of societies all over the world create abstractions of nature, the level of abstraction enacted and voiced by my informants seemed high. Since human intervention in nature (sculptured gardens, *bonsai* trees) was not seen to detract from, but rather to enhance, the beauty of natural forms in Japan (see esp. Asquith and Kalland 1997), concern in urban Japan over human transformations of "wild" settings also seemed muted relative to other societies in which I have lived long-term. Toward the end of my core Tokyo fieldwork on this project, however—after approximately 2006—the influence of environmentalist discourse on Japanese attitudes to nature and environment became much more apparent, a theme to which I return in Chapter 8.

Against the backdrop of such rarefied natural-cultural forms and abstracted relations with nature, a controversial influx of rapacious, waste-scavenging vermin in Japanese communities offended Japanese sensibilities, challenged urban ideas of ecology there, and prompted responses that help elucidate Japanese approaches to sustainability in the new millennium.

CHAPTER 5

Tokyo's Vermin Menace

Though Japanese urbanites are not "unique" in their desire for (socially construed) order in their communities, they tend to be fastidious, and their communities do tend to be orderly. It therefore comes as no surprise that an enduring blight of scavenging crows *(karasu)* appalled Tokyo residents. In a society with highly normative ideas about nature, *karasu*, swooping black birds with seemingly relentless hunger and jarring cries, became a part of Tokyo's environment that most residents would have preferred to edit out of the ecology. Significantly, the population of *karasu* had ballooned largely in proportion to the sharp rise of community waste in the high-speed growth of the 1980s, for *karasu* fed greedily on domestic waste left out for community collection each week. *Karasu* were therefore both a nuisance growing out of Tokyo's waste plight and an indication of how easily Japanese idealized constructions of "nature" and order can be subverted by elements of the archipelago's own emergent ecology. Though at first glance a community dilemma, Tokyo's government was forced to mobilize to reverse the deteriorating situation, due to *karasu*'s adaptable ingenuity (see Marzluff and Angell 2005), their occasionally violent attacks on citizens, and the public's shrill outcry regarding this conspicuous nuisance. As a consequence, this case allows us to examine the broad ramifications of a specific waste quandary and a very Japanese attempt to enforce "sustainability" through scorched-earth tactics.

Karasu were, in fact, deeply entangled in the waste predicament of Japanese cities, in the environmental engagement of Japanese urbanites, and in how Japanese translate rhetoric both over Japanese nature and over sustainability into actual conduct in their communities. These pesky, irksome, noisy fowl offer a convenient vehicle for interpreting many-valenced attitudes toward "environment," "nature," and health circulating in contemporary Japan. At the same time, the crow problem also stimulated considerable home-grown discourse on the topic of "wasteful" consumption and debates over the human role in creating and exacerbating the *karasu* epidemic.

"Harmful Birds"

Karasu (specifically *hashibuto-garasu,* or, in Latin, *Corvus macrorhynchos japonensis*), are members of a jungle crow species that appears throughout Japan and East and Southeast Asia. A somewhat smaller, less intimidating, less controversial cousin, the carrion crow *(hashiboso-garasu)* lives mainly in woods and other rural areas of Japan, foraging for seeds and fruit, feeding on small animals and carrion, and so on. While *karasu* have long made forays into human spheres, encroaching on agricultural lands and rural landfill in their wide-ranging search for food—the term for winged vermin in Japanese is *gaichō,* or "harmful bird"—in the past two decades a newly emboldened crow, adaptive to the most intensive human environments, moved more conspicuously into Japanese settlements, particularly its largest cities, like Tokyo. These "urban jungle" crows, merely annoying early on, transformed into a major public menace: creating a din in the early morning hours; roving the skies in gang-like murders while on reconnaissance; standing watch in trees, on rooftops, and on overhead wires, waiting for residents to bring out the garbage; tearing into waste collection stations with sharp beaks; and in time influencing community behavior in numerous subtle and not-so-subtle ways. Their rowdy ubiquity made *karasu* both a prominent element of everyday life in community Tokyo and a vexing component of Tokyo's warped urban ecology.

An examination of *karasu* and their impact furnishes insight not only into Japanese urban thinking about the environment, about waste, about animals, and about community order and responsibility, but it also sheds light on government attempts to enforce "sustainability." First, however, I describe Tokyo's waste regulations and community compliance, since these preceded the crow invasion and then adapted to confront the corvine threat.

Waste Protocols: Bagging, Dumping, Thinking

Incineration of waste was not some distant subject aired exclusively in the mass media; it was firmly embedded in the everyday waste protocols of Tokyo residents. When I embarked on ethnographic fieldwork in 1998, residents in my field communities separated their waste daily into "burnable" and "unburnable" categories to conform with strict government policies in the capital, and much of their discourse on waste practices was phrased in these terms. That is, rather than discussing waste *(gomi)* in general, many informants spoke about burnable waste or unburnable waste, illustrating the extent to which the terms the debate had already been set. Burnable waste *(moeru/kanen gomi)* included organic matter such as leftover food scraps, dirty diapers, and binned paper products. This class of waste was, at the time, typically collected three times a week due to the prospect of rotting garbage and the stench that emanated from such waste. Unburnable waste *(moenai/funen gomi)*, on the other hand, was usually collected once a week and often consisted of plastics and other nonorganic packaging. Ideally, unburnable waste was meant to be scrubbed and rinsed of organic residues (such as *miso* paste still clinging to the sides of its plastic container, or chicken slime lingering on a Styrofoam tray) so that it would not give off odors or attract vermin during its weekly wait for collection. But in practice this time-consuming job was not always performed to the letter. Burnable waste was destined for Japan's many incinerators, while unburnable waste tended to end up in landfill, which included land reclamation projects. In theory, the burnable waste was more or less organic and biodegradable; unburnable waste, with its plastics and synthetic materials, was not. But in fact, all sorts of chlorine-based products reside in so-called "burnable" waste that can release dioxins and other toxic chemicals into the atmosphere (see Institute of Medicine of the National Academies 2003; World Health Organization 1999; United Nations Environment Programme 1999; Casper 2003).

Perplexed by waste proliferation particularly after the Bubble economy burst in 1991, Japan made recycling a high priority by the mid-1990s, but the change was gradual. When I arrived in 1998, recycling practices were just beginning to become more organized and state-directed; these were early steps toward the aggressive, integrated recycling effort under way by 2006. In Horiuchi, there was a "resources" *(shigen)* collection day every fortnight, predominantly for old newspapers, clear plastic (PET) bottles, glass bottles (excluding beer bottles), and metal cans. Beer bottles were to be returned to the stores where they were sold and were often reused or reprocessed by the brewery companies and distributors themselves.

Stricter state measures, in turn, spawned an entrepreneurial sector of "recycle shops" *(risaikuru shoppu)* that developed partly to exploit the high mandated costs of disposal paid by households. Tighter regulations had, by then, limited the amount of *sodai gomi*, or "bulky waste," residents could jettison for state removal (collected usually once monthly and for steep fees). Unlike the mass of good-quality electronic goods and furniture that was cast onto the streets during the Bubble period's days of conspicuous consumption—and conspicuous disposal—most nonstandard, heavier, or larger-scale disposal during the time of my fieldwork was subject to expensive disposal surcharges. As a result, where possible, some residents instead sold such bulky waste to companies that had sprung up to recycle, refurbish, and resell televisions, VCRs, refrigerators, toasters, table lamps, and the like. Since the companies were cynically aware of the difficulties residents faced, their offers were extremely low in price relative to perceived value, and transactions involving large objects like refrigerators served mainly to allow sellers to avoid disposal fees as well as the chore of simply moving the appliance.

As in most Japanese residential communities, conformance with waste disposal regulations in Horiuchi was strictly monitored and controlled, largely by members of the community themselves, despite inevitable zones or constituencies of less-than-enthusiastic compliance and even resistance (see figure 6).[1] A network of small waste clusters bound neighboring households together into webs of cooperation and obligation exclusively for the execution of local waste collection protocols. These did not duplicate other political units on the micro-level of Japanese communities, past or present (see Dore 1973). My tatami-matted *hanare* residence, little more than a shack located behind a main house, fell technically within the Iwasaki household, and we belonged to a waste cluster of four roughly contiguous households that shared the responsibilities of looking after "our" waste station. Somewhat ghoulishly, our waste cluster included the cemetery caretaker and family, and the waste station itself lay just next to the cemetery, though I never witnessed dumping of anything other than household waste there. Every morning on waste collection days, we deposited our bags of refuse in the appointed area, against the tall outside wall of the cemetery. Collection ran like clockwork. One person from each household, nearly always the housewife, had the monthly, rotating obligation *(tōban)* to ensure that the rubbish heap was tidy and kept out of the way of passing traffic. Most lanes in Horiuchi were so winding and narrow that navigation by car grew challenging at the best of times; on *gomi* days, maintenance of waste stations was, therefore, essential if only to maintain traffic flow.

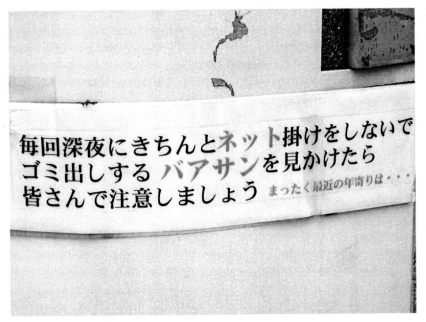

毎回深夜にきちんとネット掛けをしないで
ゴミ出しする バアサン を見かけたら
皆さんで注意しましょう まったく最近の年寄りは・・・

FIGURE 6   A sign written by an exasperated Horiuchi landlord asking residents to keep an eye out for an old lady *(bā-san)* who stubbornly dumps her waste late at night and does not cover it with the antivermin netting.

And after a team of uniformed and helmeted sanitation workers spirited the waste away via small truck, usually between 8:30 and 9:00 a.m., the waste monitor would come out almost immediately with dustpan and broom to sweep up any residue. Even if there was none, the monitor would often make a few quick sweeps of the broom to fulfill their responsibilities. Such embodied gestures conveyed the extent to which the gaze of others drove compliance with waste protocols.[2]

It is worth speculating how such a system could function without the institution of the Japanese housewife, who is at home and available to handle waste disposal duties. As time went on, and particularly as the economy deteriorated, some unemployed and retired husbands slowly became more involved in the process, though begrudgingly.[3] It was intriguing in this regard that the term "bulky waste" *(sodai gomi)* became slang used by wives to refer to husbands who used to work but were now hanging around the house getting in the way of the wives' accustomed duties and freedoms. With the economic distress of the late 1990s, and the increase in layoffs, bankruptcies, and forced or early retirements, the number of husbands no longer at work swelled, creating frictions within households. Some of these men tried to disguise the situation, according to locals, leaving at the same time in the morning wearing normal work clothes to look for a job, but the "bulky husbands" to whom I spoke made it clear that they did not like to be engaged in the very public, female-dominated, and rather demeaning activity of waste station monitoring when their household's turn came.

Though these descriptions involve key elements of the human process of waste collection, they lack the presence of the winged vermin who preyed relentlessly on poorly guarded waste stations and sometimes on waste monitors or their pets. Indeed, many of Tokyo's newer waste measures were explicitly conceived to combat the *karasu* problem.

"Vultures, Vultures Everywhere!"

Karasu are smart, I'm telling you. They remember faces. A friend of mine went out and tried to shoo one away, and the next day, it swooped down and attacked her when she came out [of her house]. It knocked off her hat even!... That's what I'm saying. They're smart, right? It's scary how smart they are.
   —Shimizu-san, *yakitori* restaurant proprietress

Karasu are scary. I've heard stories about crows attacking children, things like that. Sometimes, when there's a lot of them in the street, I'll push Ken-chan's stroller away [on a detour] rather than get closer. *Karasu* are spooky—I really hate it when they're around.
   —Suzuki-san, Horiuchi housewife

It is in the twice- or thrice-weekly rhythms of waste disposal, collection, and removal that the black, swooping, almost apocalyptic-seeming murders of crows—so common in contemporary urban Japan— make their entrance onstage in the choreography of local waste practices. The blight of scavenging *karasu*, closely linked to flows of waste in urban Japan, seemed nearly ubiquitous in many communities in the capital, a sure sign for travelers that they had arrived home from trips abroad, for example. With proximity came familiarity, though hardly the kind of affection that Japanese bestow on animals, such as pets, that conform more obediently with norms regarding appropriate interspecies behavior.

While the bureaucrats who crafted Japan's postwar waste disposal and collection were able muster a wide-ranging network of stakeholders, and while they could depend on a largely willing and obedient citizenry to execute state dicta, they did not fully anticipate the impact that winged vermin *(gaichō)* such as *karasu* could have on their careful waste plans. Small but energetic murders of jungle crows would, in time, inflict great inconvenience on communities, not to mention expense. Upon my arrival in 1998, the *karasu* problem *(karasu [no] mondai)* was becoming a major drain on resources and patience. Early in the morning each *gomi* day, these scavengers ripped through bags of waste and littered communities, prompting a national debate as to who bore responsibility and how the problem could be corrected. They also inspired a raft of clever counter-*karasu* measures *(bō-karasu-taisaku)* that drew on ornithological insights.

Most Tokyo residents I knew were excellent sleepers, but *karasu* typically created a rousing din when they descended on waste stations. Hearing the thrumming wingbeats and shrill caws of *karasu* dopplering past one's window was a reliable reminder that *gomi* day had arrived.

Locals found the *karasu* nuisance visibly perplexing. Waste regulations mandated the use of translucent bags in order to discourage illegal dumping. While this translucence was a bulwark against expensive lapses in sorting of discards, it also helped *karasu* locate scraps. The crows tended to be very good at finding the gobbets they sought—meat scraps being their main prize—and could quickly peck through the bags to get their breakfast. The act of doing so, however, left waste stations in disarray, for the crows tore and sometimes destroyed the plastic bags, strewing waste about. With space at a premium in urban Japan, this mess could create a real eyesore, as well as circulation problems; many community waste stations sit directly in the street, there being no sidewalk.

When I arrived in Horiuchi, several rudimentary antivermin measures were in place, but these were flawed. For example, nearly every Tokyo waste station was outfitted with green or blue plastic netting to cover the bags of waste; it was specifically designed to be cheap, easy, durable, and effective against the *karasu* menace (see figure 7). Residents spread this webbing over their bags of waste or lifted the mesh and tossed the bags under it.[4] At many waste stations, monitors and community leaders posted signs emblazoned with cartoonish images of *karasu* and warnings against improper disposal, including lax use of the green mesh. Some of these were stark, brightly colored reflective signs that recalled "biohazard" warnings. Others made use of the black, sinister aspect of jungle crows to make the notices more striking or memorable (see Figure 8).

FIGURE 7 A Horiuchi waste station with antivermin netting. Note the many signs urging correct disposal practice and warning against illegal dumping, including the antisocial dumping of dog droppings in other people's waste stations.

FIGURE 8 A sign, bordered in reflective orange, urging Horiuchi residents to exercise caution *(chūi)* in covering their garbage bags with the antivermin netting.

The voraciousness and ingenuity of *karasu* made the timing of disposal crucial. Whereas waste disposal in many world capitals can be rather laissez-faire, with residents dumping their waste in covered garbage cans, and even putting it out the night before collection or frequently even earlier, Tokyo waste monitors, community leaders, and the bureaucrats who dictate waste policy insist that waste goes out literally just before collection. Less time out in the liminal space of the waste station means less time for vermin of any kind—including rats, but especially *karasu*—to rip into waste-heaps (see figure 9). This requirement, however, creates more trouble for residents. "Just-in-time" disposal meant that households and caretakers were forced to store their waste indoors for longer, which sharpened resentments with elements of the capital's waste policy, as well as criticism of local practices. For example, Iwasaki-san, my landlady, complained that the next community over from Horiuchi—more urban in feel, with some office buildings—used plastic garbage cans, which made life much easier for waste monitors. "Why can't we use bins like they do? Our *chōkai* won't allow it, but it makes no sense to me.... That way, there'd be no *karasu* problem at all." She did admit, though, that the neighboring community was colder,

FIGURE 9   A sign warning residents not to deposit
their garbage too early, due to vermin.

with far less of a community ethos than Horiuchi. "It's true," she confessed. "It's better to work together." Residents who lived in apartment complexes or in more upscale *manshon* often had a ground-floor room with a door where they could deposit waste safe from crows and other vermin, but illegal dumping was reportedly more of a problem in such places, and microcommunity solidarity was palpably low in comparison to our neighborhood with its fastidious waste clusters.

With *karasu*, even cooperative bastions of relative solidarity like Horiuchi had their work cut out for them. Over time, as the population of *karasu* reportedly increased, contact with humans increased as well, and the low-level psychological trauma *(seishinteki [na] higai)* caused by these shrill vermin deepened. *Karasu*, according to locals, seemed to engage in calculated attacks against aggressive waste monitors. They intimidated pets in yards and on walks, disturbing both the animals and their owners. In the spring, particularly around breeding season between April and June, parent crows protective of their young became territorial. The birds retaliated against intrusions sometimes through intimidation and sometimes through physical attacks with their sharp beaks or talons, drawing blood on occasion. These assaults on humans attracted a great deal of attention toward the late 1990s and constituted a clear impetus, or pretext, for action; authorities finally decided that the *karasu* population, once merely an annoying ecological distortion, had developed into a menace. Tokyo Governor Ishihara Shintarō's declaration of "war" against *karasu* during his first term in office, in 2001 (Okuyama 2003), ensured that *karasu* would remain prominent in Tokyo policy throughout my fieldwork. The governor's campaign, though for wider benefit, was also personal: anecdotally, according to city officials, Governor Ishihara angrily ordered the capital into action after a jungle crow buzzed his head during a golf game.

Before taking up elements of Governor Ishihara's many-pronged "war" on *karasu*, however, I first want to show how the *karasu* problem was intertwined with regulatory choices, with residents' own waste habits, and with Japanese society's largely enthusiastic postwar embrace of mass consumption.

Ecology Close to Home

Though the Hitchcockian pestilence of crows in Japanese communities might seem like a mere ornithological curiosity, or an ironic sign of the times, *karasu* were inextricable from grave waste issues besetting Tokyo. From 1985 to 1989, waste levels swelled in the twenty-three wards of Tokyo to well over 5 million tons, an increase of 1.1 million tons (e.g., Ueta et al. 2003). Correspondingly, from 1985 to 1999 the population of *karasu* rose threefold, to 21,000, according to the Tokyo Metropolitan Government, and an abrupt increase appears to have occurred in the final few years of that period. From a reported high of 36,400 crows in 2001, Tokyo claimed to have reduced the population to 17,900 by 2005 (Tokyo Bureau of Environment 2006a; 2006b, 107). It beggars belief that a shifting, mobile population such as that of urban crows could be counted with great accuracy, and the round numbers used at regular intervals (7,000 in 1985; 14,000 in 1996; 21,000 in 1999) before Tokyo's campaign intensified are highly suspicious, but the sharp increase matches the impressions of my informants. Ornithologists and other analysts unaffiliated with the government estimated a crafty and tenacious crow population as high as 150,000 in the capital (see Matsuda 2000, 2006; Karasawa 2003; *New York Times* 7 May 2008).

While reactions in my fieldsites to the growing plague of *karasu* were certainly negative in tone, some urbanites, particularly from the mid-1990s, began to question whether it was the crows who were to blame or rather the profligate, unsustainable ways of Tokyo residents themselves. This dimension of self-critique took several forms. First, there was the moral level, at which a conspicuous and pervasive waste-related problem, the *karasu* nuisance, prompted extensive debate over the excesses of consumption-obsessed, fragmented urban Japanese society. Several series in the *Asahi Shimbun* between 1998 and 2000 in particular (e.g., *Asahi Shimbun* 6 August 1999; *Asahi Shimbun* 7 August 1999), and numerous opinion pieces and articles in the major dailies more generally, used the *karasu* problem as a platform to question the moral compass of Tokyo residents. And while Japanese terms[5] lack the double meaning of the word "waste" in English—that is, the dual inflections of discarded residue on the one hand and immoral excess on the other—this level of debate transformed into a wider social critique. It both compared contemporary society to retrospective constructions of earlier, thriftier, and more virtuous (and, not coincidentally, "traditional") generations and played off of pre- and postwar campaigns for frugality (Garon 1997; Havens 1978; Dower 1986, 1999).

During my initial fifteen months in Tokyo in 1998–1999, my informants expressed frustration over the gloomy economic climate, the high unemployment rate, and the abrupt rise in bankruptcies, as well as the breakdown of certain perceived cornerstones of Japanese social order, such as the strong, affective ties of loyalty and familial interdependence within Japanese companies. These topics prompted some soul-search-ing with regard to the course that Japanese society had taken and the occasional lack of community cohesion and support evident to some during these difficult times. Tokyo informants pointed to conspicuous consumption, but especially to conspicuous waste and defilement of nature, as elements of Japan's shift in values. A comment during a wide-ranging discussion of nature, waste, and *karasu* was emblematic: one evening a single mother in her thirties, resident in the community for six years, muttered into her *mugi shōchū* (distilled barley liquor) in a local bar: "Japan is sick *(yanderu ne)*. No question about it *(machigai nai)*. But no one knows what to do about it." The rise of the *karasu* nuisance in Japanese communities like Horiuchi made the situation seem more problematic, as if "nature" itself was chastising the direction Japanese society had taken.

Next, there was the way in which the *karasu* problem became dis-cursively intertwined with rhetoric over sustainability. During the 1990s, particularly after the 1992 UN summit on sustainable development in Rio de Janeiro and subsequent negotiations in Japan over what came to be known as the Kyoto Protocol (the UN Framework Convention on Climate Change), adopted in 1997, Japanese awareness of sustainabil-ity and other activist priorities of global environmentalism expanded noticeably. It began to dawn on Japanese communities that they were not existing in sustainable balance with their wider ecologies, a realiza-tion mirrored in the growth in volunteer recycling activities and other waste-related attempts to create a "sustainable society" *(junkan-gata shakai)* as well as in the increasingly energetic bureaucratic efforts to control Tokyo's growing waste disposal predicament. Echoing a line of thought aired by a number of other outspoken residents, Furuhashi-san stated, "[It's all] out of balance. We're out of balance with nature. Anyone can see that." Later, responding to a question specifically about vermin crows, she went on to voice a quasi-superstitious comment about crows[6] that I had heard before, but this time with an environmentalist slant: "It's like *karasu* are a bad omen *(warui engi)*.... We Japanese bear a big responsibility for what we've done, and it'll probably get worse."

The entwinement of *karasu* in the discourse on sustainability was significant. For Tokyo residents *karasu* were an unusual and persistent reminder that they did in fact live in an "ecology" of sorts and that their conduct did have repercussions within their immediate surroundings. These beliefs bled into discussions of the *karasu* problem and its origins. One female restaurateur commented: "*Karasu* came down to the cities because there wasn't any more room in the mountains. [There's so much] cutting and quarrying and construction, and food started to run out. So they came to the big city *(tokai)*. Now that they're here, they've got plenty of food." A customer in the restaurant in his thirties echoed this line of thinking: "It can't be helped, can it. The mountains are disappearing *(yama wa nakunacchau)*. Where else can they go?" This mention of sylvan scarcity, while not uncommon, was strictly inaccurate. Japan offers surprisingly abundant woodlands (though sometimes man-made, planted with industrial cedar and the like) and other more or less suitable habitats for birds not lured into human settlements.[7] While these two informants, and others, did not emphasize waste as cause rather than effect, their later comments made it clear that they believed residents of Tokyo produced too much waste. An elderly woman who was a long-time resident was quicker to draw clear connections between the material and moral dimensions of waste: "*Karasu* thrive because there's so much to eat in *gomi* here," she said. "You wouldn't believe what people throw away these days!... We used to take chicken bones and vegetable scraps and make broth [out of it all]. With leftover rice, you make *onigiri* (rice balls) or *zōsui* (rice gruel), any number of things. But people toss vegetables now because they think they're old. Some don't even know *how* to make soup!" If we consider the "ontology of detachment" (Ingold 1993, 41) that can result from a focus on global environmentalist themes to the exclusion of local environmental engagement, embroilment in the *karasu* nuisance certainly led to an experience of the downside of unsustainable practices that went beyond mere abstract rhetoric.

Karasu did have their defenders, though they remained a minority. Japanese who were more vocal in condemning human before crow tended to have a scientific or naturalist background (from high-school or university education, for instance, or from club or association activities). They viewed *karasu* scavenging as a simple response to ecological influences, most of which were the result of, or at least under the control of, human society. More sympathy was voiced for *karasu* after the Tokyo government launched a multipronged campaign against the birds that included the intensive culling of crow eggs. Newspaper and television reports of distraught parent crows fighting to prevent city workers from abducting their young struck a chord with residents; the crows' apparent parental concern and ties of affect seemed almost to humanize these pests and bolstered sentiments that sought to assign blame for *karasu* overpopulation to unsustainable human practices. However, I found not a single informant who thought the crows were likeable, let alone members of the menagerie of creatures some Japanese consider so cute, cuddly, or adorable ("Hello Kitty").[8]

Japanese Sustainability with an Iron Fist

Governor Ishihara's belligerent attitude toward crows manifested in a sustained offensive against vermin in Tokyo's waste policy. Due to the capital's central role in Japanese life historically as well as its serious waste predicament, Tokyo's campaign against *karasu* became a strategic blueprint for the rest of the nation. Innovations from beleaguered towns and cities in the provinces, in turn, influenced strategy in the megalopolis.[9] Some of Tokyo's efforts included tightening up elements of the waste collection routine—which were fundamental—but most notable was the capital's close ornithological attention to jungle crow behavior both inside and outside urban confines. Examining the crows' engagement with features of the urban ecology furnished metropolitan officials with ammunition to combat the *karasu* nuisance. Essentially, bureaucrats and their minions began to use the crows' "natures" against them.

Insights gleaned from naturalist study helped the authorities gain ground against the *karasu* menace. Take, for example, the timing of waste collection. Waste monitors had long been urged to make sure waste was put out as close to collection time as possible. But this was hardly foolproof, as crows could still lift an ill-fitted plastic net and even peck through or around the netting. The problem was even more serious in entertainment districts like Shinjuku, Shibuya, and the Ginza, where the leavings from bars and restaurants might be put out carelessly by wage-earning nonresidents and sometimes hours earlier than recommended. Officials hit upon an ornithologically inspired solution: collection could take place at night, when crows do not ordinarily feed. Nighttime collection *(yakan shūshū)* was approximately 50 percent more costly than daytime collection, it created noise that disturbed sleeping humans, and it required some other adjustments on the part of communities—for example, adjusting routines to the much later disposal times for waste clusters. But residents and officials deemed nighttime collection very successful when it was piloted in the Tokyo suburb of Mitaka City in 1998 (e.g., *Japan Times* 2 July 2000), and it was then expanded in 2000. Nighttime collection was particularly well suited to the entertainment districts. Tokyo had discovered that knowledge of the rules of the natural world could strengthen its hand in improving waste policy.

Rigorous investigation of crow anatomy yielded further insights. For example, Utsunomiya University scientist Sugita Shōei determined that *karasu* searched for food not by smell but by sight. According to Sugita, superior crow eyesight (approximately three times sharper than the human eye) could be thwarted by using specially prepared translucent yellow garbage bags. To the crow these bags appear opaque, but humans can see through them, translucence still being deemed important by Japanese authorities for monitoring proper separation of waste (*Yomiuri Shimbun* 10 March 2006; *Tokyo Shimbun* 5 December 2005; see also Sugita 2004). Tokyo's Suginami Ward began experimenting with the bags in 2005. Sugita's claims regarding the opacity of the yellow trash bags to crows have not yet been properly verified by other scientists, partly due to proprietary discretion regarding his formula, and according to a Tokyo bureaucrat tasked with the *karasu* problem, clever crows had by 2009 found their way around this putative yellow barrier.

Tokyo's government also launched a comprehensive information campaign to educate the public. While urging residents to dispose of their waste correctly according to regulations, the campaign informed them of the behavioral and ecological roots of crow conduct. For example, the campaign explained that attacks on humans usually occur during the breeding season between April and June, when parent crows become territorial over threats to their young (Nature Conservation Bureau 2001). Campaign materials showed diagrams of how crow habits of perching on high tree limbs in the forest to look for food resembled their use of rooftops and electrical wires as vantages from which to reconnoiter vulnerable waste stations in cities.[10] Throughout the materials, the campaign explained the ecological influences on crow behavior and the human causes of the crow nuisance (e.g., Nature Conservation Bureau 2001).

Yet this kindly public face of *karasu* policy was combined with a ruthless, multipronged culling campaign. In their annual census of *karasu*, city officials identified key *karasu* roosts in Yoyogi Park and in the huge, sylvan Meiji Shrine compound, among other locations. In wooded parks and elsewhere, functionaries set traps to cull adult crows. One effective, widely used method placed bait (scraps of meat, usually) in plastic bags filled with poisonous gas; when crows pecked through in search of food, they would be killed by the fumes. Over approximately four years, until March 2006, Tokyo culled 67,000 live crows—almost twice the claimed population of crows in the capital (Tokyo Bureau of Environment 2006a; 2006b, 107). Workers combined this approach with

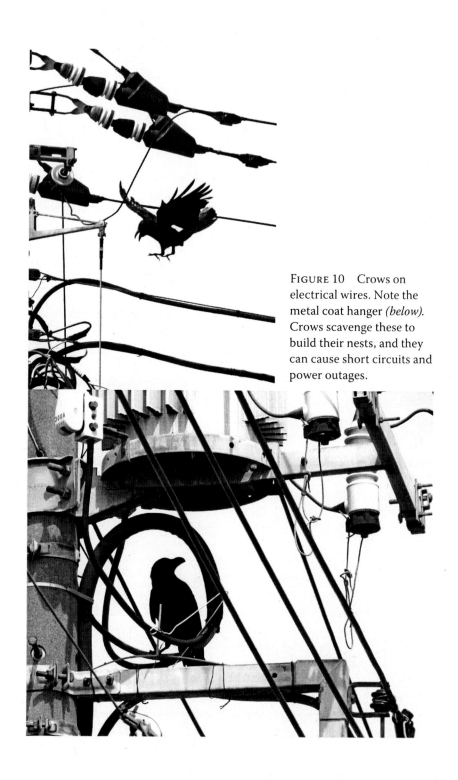

FIGURE 10　Crows on electrical wires. Note the metal coat hanger *(below)*. Crows scavenge these to build their nests, and they can cause short circuits and power outages.

an aggressive culling initiative during breeding season in which they collected tens of thousands of unhatched eggs. Since this brought human and crow into acrimonious relation, cullers donned armored gloves, helmets, face masks, and other protective equipment to protect against frequent attacks by distraught birds, sometimes in front of television cameras whose programs broadcast such clips in my fieldsites.

There was something of a mixed message transmitted to Tokyo-ites, then, as the flamboyant Governor Ishihara continued to swagger in the glare of mass media coverage and issue aggressive statements with reference to *karasu*. In his first term in office, Ishihara referenced Charles Dickens in stating that Japanese should learn to eat crow pie, a dish he said was well liked by the English (e.g., *New York Times* 30 March 2000). If Japanese made crow pie into a local delicacy, he reasoned, there would *be* no more *karasu* problem.[11] His propaganda crusade against Tokyo's voracious corvine hordes presented a brusque approach to environment and sustainability that was, for the most part, short on sensitivity or nuance. But in tandem with this the ministries and agencies tasked with the crow problem transmitted a message that emphasized understanding of the crows' relationship with their environment and the ability to live together in harmony (e.g., *New York Times* 7 May 2008; Nature Conservation Bureau 2001). Government efforts to encourage sustainability drew on both such discourses in order to persuade Japanese to become more conscious of their waste.

Karasu scavenge not only for food but for materials with which to build their nests. While all sorts of plastic scraps and twigs are found in their nests, *karasu* preferred scraps of wire and coat hangers, which they tend to purloin from balconies where household laundry is hung. *Karasu* use so much metal in their nests, in fact, that this can create serious problems with the electrical infrastructure—coat hangers short-circuit exposed wires when nests are built on electricity poles (see figure 10). They also damage the telecommunications infrastructure, for crows sometimes bite through fiber-optic cables. But the greatest threat *karasu* posed was to Japanese ideas of order in communities, challenging Japanese ways of conceiving of nature and surroundings.

CHAPTER 6

Pure Obsession
Pollution, Outcasts, and Exclusion

> [H]omeless people formed an encampment in the val-
> ley of high-rise commercial and business buildings
> near JR Shinjuku station....The [Tokyo Metropolitan
> Government] responded by sweeping the encamp-
> ment....Construction officials conducted the sweep,
> saying it was "removal of street garbage."
> —M. Hasegawa, *We Are Not Garbage!*

The conceit of Japanese homogeneity finds considerable traction among Japanese in a rather inwardly oriented nation, but it is nevertheless an illusion (Tsuda 2003; Davis 2000; Eades, Gill, and Yamashita 2000; Dale 1986; Befu 1993, 2001). One characteristic of this ideological chimera is that the normative spin put on emblematically Japanese traits by conservative elements in Japanese society leaves out numerous social groups in Japan that do not meet rigid (if largely unspoken) membership criteria. Belonging and its pursuit drive much of Japanese social life—probably even more than in most other complex, developed societies—and boundary maintenance along the margins of social groups can involve directed, even vicious, exclusion of those who do not fit. In this chapter I examine operations of social exclusion in Japan—processes rarely fully acknowledged but very much present in this group-fixated sphere—and trace the extent to which the grammar, as it were, of exclusion has some basis in ideas of pollution that pervade Japan's wastescape and influence social relations there.

In a society where status and group membership are paramount, social exclusion looms as a base, ugly counterpoint to the lofty meritocratic ideals of the Japanese postwar system and the expansive middle class it has engendered (e.g., Vogel 1963). Some of this social exclusion gained impetus from ideas of pollution and difference that have a long history in Japan. The long-shunned *burakumin* outcaste—Japan's "untouchables"—and other cruelly marginalized groups, such as a large population of Koreans, born in Japan to Japan-born Korean parents but denied full rights of citizenship, are well known to observers of Japan and bear similarities to disenfranchised groups in other societies, though with certain distinctive features. Less well known, however, are the cases of "mainstream" Japanese who fall ill, due to environmental exposure, contagion, or other misfortune, and who then find themselves ostracized by the very society that earlier embraced them. Though a broad and varied range of victims of this exclusionary process appear throughout contemporary Japanese history, including postwar lepers, the homeless, the handicapped, foreign guest workers, and even casualties of the sarin nerve-gas attacks on the Tokyo subway in 1995, I focus here in particular on victims of toxic pollution and other waste problems, the ways that ostracism of these victims bore a specifically Japanese character, and the manner in which victims responded to their newfound (though often not terminal) status. The *social pollution*, as I call it, that is perceived by Japanese groups as emanating from such designated, socially construed pariahs bears an important relation with both contemporary environmental pollution and conceptions of ritual pollution that linger on from Japan's social, religiospiritual past, and so the chapter places this ostracism in the wider context of exclusionary thought and action in Japan.

One key trigger of this exclusionary impulse is the notion of purity.[1] Purity has particular resonance in Japanese history and culture; it is positioned as a defining trait of Japanese superiority and difference vis-à-vis tainted others. But purity reverberates throughout every society, and inquiry into its reckoning by anthropologists and others can shed light on important elements of "Japanese" purity and defilement that lurk not just in the vestiges of Japan's Shinto-animist legacy but in a range of contemporary practices.

In these pages I first consider how anthropological analyses of purity and pollution can contribute to the interpretation of contemporary Japanese identity and exclusionary practices and then travel to my fieldsites in Tokyo to excavate sometimes deeply embedded notions of purity, contamination, and "untouchability" as they lurked beneath the charged sociopolitical terrain of community environmental pollution disputes and toxic anxieties. By delving into the valences of purity and the politics of exclusion in contemporary Japan (including their links to traditional Japanese constructions of hierarchy and hygiene), I hope to deepen anthropological understanding of environmental thinking and social action in Japan—social phenomena that in many ways demonstrate the centrality of purity and defilement in varied social contexts in contemporary life.

Purity of Motive

Ideas of purity have long influenced the structure of social life, creating nuanced divisions between the repulsive and the soothing, the perilous and the secure, the contaminated and the clean. The well-known work of Emile Durkheim (1995 [1912]), for example, convincingly identifies the importance of the "sacred" and the "profane" in ritual action and in mundane spheres in primitive societies, findings that certainly apply to other societies. Van Gennep (1965 [1908]) and later Turner (1967, 1969) engage with the potential pollution of liminal states and address the role that purification can play in managing movement through phases of rites of passage (though purity is not a strong focus of their work). It is Mary Douglas' (1966) seminal work *Purity and Danger* in particular that demonstrates how maintenance of the pure and revulsion at the polluted are not limited to primitive, classical, or "exotic" non-Western societies. While purity resonates differently in disenchanted contemporary settings less preoccupied with divine grace, spiritual cleanliness, and ritual conduct than in preindustrial settings (for a classical example of this, see Parker 1983), ideas of purity continue to influence behavior in many parts of the postindustrial, even secularized world and diffuse into such diverse arenas as hygiene, architecture, immigration, civil engineering, sanitation, public health, and—significantly—environmental conservation and activism. Examples of preoccupations (or, at times, strategic amnesia) with regard to purity surface predictably in discussions of "race," blood, ethnicity, and nationalism (e.g., Malkki 1995; A. Smith 1992, 2000; Gellner 1983; Geertz 1963; Conner 1978), as well as in analyses of caste, ritual pollution, and classification, particularly on the vertically configured Indian subcontinent (Dumont 1980 [1967]; Srinivas 1978 [1962]; Deliége 1999). But the salience of ideas of purity, hygiene, and social pollution as embroiled in environmental issues remains a relatively underexamined arena of investigation.

Purity has enjoyed an especially central place in Japanese society and its self-imaginings. Ohnuki-Tierney (1984, 35), one of the few scholars to give the topic the subtle attention it deserves, concludes that "purity and impurity notions constitute not only a significant principle providing order to the world view of the Japanese, but also an important part of their ethos, to which Japanese relate and react with the strongest of emotions." Pollution avoidance in outlying Japanese communities, such as mountain villages or fishing communities, bears a fairly close relation to concerns over purity and pollution in other less "developed" rural contexts in other parts of the world, with important cultural variations (see Namihira 1977, 1984, 1985; Kalland and Moeran 1992; Miyata 1977, 1996; Sakurai 1970, 1988). Furthermore, ideas of purity inform wider Japanese national-cultural myths regarding "racial" origins, distinctiveness vis-à-vis tainted foreign elements, and prewar and wartime rationales for invasion and colonization of East Asia (e.g., Young 1998; Havens 1978; Dower 1986; cf. Wendelken 2000), reservoirs of national spirit amid wartime deprivation (Benedict 1947; Dinmore 2006), and so on. Purity became not just a state to maintain during ritual action or a flashpoint in terms of community welfare (e.g., avoidance of low-status intermarriage or miscegenation) but rather the rationale for the state in toto, the very source of strength, vitality, and difference that distinguished the Japanese as the chosen people. I engage first with the larger-scale impact of Japanese purity beliefs, which shaped Japanese contemporary life through important societywide geopolitical stances in the twentieth century, before probing into the roots of this purity consciousness in everyday life in Japanese society.

One important division between pure and defiled, largely maintained at the imperial frontier, existed between Japan and the foreign. During Japan's pre–World War II colonial period there emerged politically convenient discourses of multiculturalism and inclusiveness sometimes referred to as "mixed nation theory" *(kongō minzokuron)* (Askew 2002, 79; Askew 2004; Oguma 2002)—the better to appropriate territory and populations in East and Southeast Asia. But much prewar and wartime rhetoric positioned the Japanese as pure vis-à-vis their barbaric foes and subjects and therefore able to draw on deeper wells of spirit and strength. Colonial propaganda and prewar claims of overcrowding and stagnancy notwithstanding, the Japanese homeland's procreative base was, by contrast, far more likely to be characterized as pure and vulnerably so (see Chapter 7).

The war's outcome forced Japan to recast its self-image of triumphant racial supremacy, and the nation retreated back within its early Meiji-period boundaries to engage in reconstruction and self-reflection. Partly in response to the humiliation of defeat and postwar occupation, Japanese elites proceeded to root among pre-1945 ideological scraps to cobble together an elaborate discourse of distinctiveness—familiar to all Japan scholars—in which the purity of Japanese blood, the unadulterated genealogy of the imperial family, the enduring crucible of Japanese nature and "climate," and other elements combined to forge an unalloyed space of cultural inclusiveness, congruent with the territory of the Japanese archipelago, that risked contamination from outside (Dale 1986; Befu 1993, 2001; Yoshino 1992).

That this approach to purity became easily distinguishable in this historical moment is not to say that attitudes toward purity and pollution sprang exclusively out of the trauma of defeat or the militarist decades of exuberance that preceded it. Nor was the boundary between purity and pollution exclusively limned along the colonial frontier at its furthest wartime expansion or along national borders following the Allied Potsdam Declaration of 26 July 1945. There were further sources of social contamination lurking within Japanese society, and we have only to think of the more or less enduring stigma that has adhered to regions and occupations associated with the traditional "untouchable" outcaste, the *hisabetsu burakumin,* not to mention the perceived horror of defilement through intermarriage. *Buraku* is the Japanese word for "hamlet," and this name for untouchables derives from their erstwhile sequestration on the margins of society in *tokushu buraku,* or special hamlets, handled as polluting elements kept at a remove (Davis 2000). *Burakumin,* technically existing outside the rigid hierarchies of traditional Japanese society—though integral to its functioning (see Ooms 1996; Ohnuki-Tierney 1987; Weiner 1997)—were deemed polluted because they customarily engaged in prohibited activities related to death such as slaughter of animals, tanning, and leatherwork (deriving from the ban against killing animals in Buddhist scripture), grave-digging and corpse-handling (seen as highly polluting in Japan), and other "perilous" tasks (Davis 2000; DeVos and Wagatsuma 1966; Sumitomo and Itahara 1998; Ninomiya 1933).[2]

There is an intriguing tension and opposition between contemporary Japanese invocations of *furusato*—the affective, inclusive, idealized repository of nostalgia for the "old village" and its traditional ways—and the inherently Bantustan-like segregation and discrimination of the (now geographically unmoored) *buraku.* It is revealing that when many Japanese speak about *burakumin,* they invoke the "Dōwa Problem" *(dōwa mondai).* *Dōwa,* meaning "integration," is a broadly used euphemism for *burakumin* whose meaning may seem puzzlingly optimistic to foreign interlocutors until its common pairing with the word "problem" is taken into account. The Dōwa Problem is, simply put, the "problem" of allow-ing these stigmatized outcasts full rights and freedom of movement. For those more bigoted against these outcasts, the "problem" is keeping *burakumin* out of mainstream life and away from non-*buraku* families in these more mobile times—by now a futile undertaking, but still com-pelling for some purity-fixated Japanese. In postwar Japan, many *bura-kumin* have moved far from their former ghetto-like communities and far from their prescribed occupations (Davis 2000; Bondy 2005). Yet the stigma of *buraku* association persists (e.g., Matsushita 2002), and though it remains illegal, *burakumin* have faced intense discrimination in, for example, finding work in high-status Japanese corporations; they have also faced towering hurdles obstructing marriage to non-*burakumin* Japanese. In both cases, parties have aggressively employed investiga-tors to dig into the background of job applicants or suitors.[3] The remote, stigmatized hamlet of the *buraku,* while geographically outdated, hints metaphorically at a roving marginalization from which there is little escape, save through careful anonymity. Many supposed "untouch-ables" circulate throughout Japanese society in this way, either hiding from rigorous scrutiny of their background or unaware of their stigma-tized heritage. Other *burakumin* embraced their status and have been employed in local institutions or businesses in state-subsidized com-munities traditionally populated with *burakumin* or have found work alongside guest workers and other "pariahs" in low-status 3-K[4] occupa-tions from which there was little prospect of advancement (Tsuda 2003; Gill 2001; Fowler 1996). While the political vectors of prejudice have changed for untouchables, and they can now enjoy both relative benefits of state support and enormous power to denounce publicly those who mistreat them, outcaste discrimination certainly endures in the present day. However, partly since the official *buraku* preference in recent years has been to stress Japanese homogeneity, a strategic stance emphasizing *burakumin* inclusion in that popular Japanese racial conception (Davis 2000; cf. Befu 2001), exact population figures are nearly impossible to compile or verify. Nevertheless, despite diffusion away from traditional outcaste trades, significant numbers of *burakumin* appear to work either directly or indirectly in waste handling, another polluting and low-sta-tus 3-K activity, as well as linked activities such as criminally organized illegal dumping and trade in pilfered recyclables.[5] Though not nearly so blatant as during Japan's early modern period, when "the dominant Japa-nese view of the special status people as 'naturally' impure and hence morally inferior was firmly established" (Ohnuki-Tierney 1987, 100), *burakumin* discrimination has not entirely subsided, either; insults and even outright bigotry occasionally surfaced in conversations I had with my informants as they grappled with the pollution (real or perceived) of waste and waste facilities and other heated topics.

Burakumin and other long-shunned groups are, in many ways, an extreme case, and they have drawn a disproportionate amount of scholarly attention. An important complementary approach, I believe, is to identify more fluid, emergent categories of contemporary "untouchability" that bear a closer resemblance to the messy realities of stigmatization and exclusion in everyday Japanese life. Nevertheless, pollution beliefs among my informants that surfaced with reference to *burakumin* and other low-status social groups hint at the degree to which Japanese society remains preoccupied with what I call "social pollution." I focus on elements of pollution as embedded in customary and contemporary practices in order to demonstrate the pervasiveness of pollution beliefs and their impact on wider present-day concerns, including social exclusion, environmental welfare, and waste broadly defined.

A Polluted World

Purity and its iconography engulf contemporary everyday Japan. The ubiquity and emotive force for Japanese of mass-mediated symbolic elements in Japan's natural-cultural repertoire suggest the extent to which this concern over purity endures in Japanese social life. Key images include cherry blossoms, which in signifying purity can symbolize life, love, courtship, rice, fragile beauty, ephemerality, unspoiled youth, nubile sexuality, and reproductive power (e.g., Ohnuki-Tierney 2002); Mt. Fuji, representing the perceived essential majesty of nature and a trove of purity remote from human failings (Berque 1997a);[6] and other symbols such as the highly evocative resonance of the seasons and the poetic change they bring—including cherry blossoms, of course, but also heaping drifts of powder-light snow, or pied autumnal foliage. Such natural symbols are frequently combined for effect—a pristine first snowfall on Mt. Fuji, Mt. Fuji glimpsed through an autumnal landscape, and so on.

It is significant, furthermore, how such natural-cultural symbols are appropriated to achieve ends, with the visual purity of mountain views or wooded expanses conveying the supposed purity of motive of banks, insurance companies, and other cutthroat commercial operations. A particular concentration of natural symbols appears in conjunction with hygienic products. The deluge of advertisements for toilet cleaners, laundry detergent, wet-wipes, dish soap, mouthwash, roach killer, and so on may suggest relative congruence with tidying and hygienic practices in other industrialized societies in that the products are more or less the same. But Japanese society's predilection for cleanliness and pursuit of control takes culturally specific forms. It comes embellished with a distinctly Japanese symbolic vocabulary whose clear links to Japanese ideas of nature are of significance to this study. To choose from among many examples: a popular toilet cleanser promoted on TV called Toire-Senjōchū (from Kobayashi Pharmaceuticals), purported to combat the filth that can accumulate back out of sight in the toilet drainpipe. The ad promised that the product "also [gets rid of] filth in the drainpipe and defends against unpleasant odors *(haisui-paipu no yogore mo otoshi, akushū no hassei wo fusegimasu),*" the visual elimination of which comprised a significant portion of the short ad. Many societies would likely agree that it is important to clean the visible parts of the commode, but this product played upon Japanese sensitivities to hidden impurities by creating anxiety over an invisible build-up of filth, the alleged cleaning of which would be impossible to verify, unless the consumer also happened to be a plumber.

Such marketing is merely one indication of a set of commodified purity concerns that are reflected in wider social behavior. To take just one example: most outside observers regard Japan's gas stations, where self-service is the overwhelming exception rather than the rule, as an inefficient anachronism. But in a society that prizes automobiles and has a low tolerance for certain forms of "dirt," the stations' salaried grifters go a long way toward paying for themselves with their practiced expertise in playing on Japanese abhorrence of impurities. Amid a great deal of bowing and obsequious service, attendants make it their business to check out the vehicle, particularly under the hood, in order to identify elements in "need" of replacement. As the team goes to work on a car during a routine fill-up, they may take the proactive step of removing an air filter rather than merely explaining the situation to the customer; when shown an otherwise functional filter with a layer of filth—there from doing its job of filtering—more than a few Japanese will find themselves bamboozled into purchasing a replacement prematurely. Attendants also regularly siphon out liquids such as oil, transmission fluid, or radiator fluid to place in ampoules next to fresh specimens. They then argue that the comparatively "dirty" fluid has reached a "dangerous" level of contamination, needing urgent replacement. While this need for replacement clearly sometimes exists, other times attendants in my experience exaggerated impurities shamelessly to clinch the sale, even criticizing fluid they had replaced just a handful of months earlier in a car used only moderately. It goes without saying that a broad array of car wash options are promoted each visit.

Another linked dimension of Japanese hygienic concern surfaced in discourse by consumers—with a great deal of repetition through product marketing—on the imperative to "refresh" *(rifuresshu)*. A host of products such as laundry detergents, bathroom cleansers, futon hanging-clips, and the like touted their utility in transforming the dank or the stagnant, mobilizing an idea of refreshening that exhibited distinct moral overtones. One informant, a Japanese university student named Kaori, used "refresh" several times in the same conversation but with different nuances that express the term's plasticity. During one interview, we began discussing how the Japanese frequently hang their futons out to air in sunlight, and I mentioned that households in Europe and North America rarely (if ever) air their mattresses. Puzzled, she asked, "Then how do they refresh them?" If her comments remained relatively polite, her appalled facial expression showed exactly what she thought about foreigners who slept on—in her mind, utterly filthy—unaired bedding. Ohnuki-Tierney (1984) writes that one Japanese phrase for a lazy person is a *mannen doko,* literally a futon left on the floor for "ten thousand years." Minutes later, while describing laundering practices, Kaori spoke about the need to dry-clean her wool sweaters and other delicate fabrics regularly to "refresh" these items of clothing, even though she confessed she was sometimes remiss in doing so. Approximately an hour later, talking about an upcoming holiday to Canada, Kaori described how much she needed the time away to "refresh" after a difficult period of study and job-hunting. All of these usages appeared regularly in Japanese discourse in my conversations and interviews, demonstrating a subtle, if persistent, concern in everyday Japanese lives with poor hygiene, stagnancy, and their alleviation. It is significant that many invocations of *rifuresshu* involved airing domestic items in "natural" sunlight, which is widely perceived to be a reliable and essential source of purification. Ohnuki-Tierney (1984) describes how Japanese would put used items like second-hand books out for a few hours in the sun to kill "germs" from the hands of strangers, and Berque (1997b, 13) stresses the importance of household access to sunlight for this purpose, a necessity enshrined in Japanese law as the "'right to sun'" *(nisshō-ken)*. Playing off of such moral nuances, *rifuresshu* also surfaced predictably in urbanites' descriptions of breaks, trips, company outings, and so on, as they left work and cramped urban conditions for brief immersion in "nature"—themed or otherwise—though the term could be used for breaks in other contexts as well.

This societywide preoccupation with pollution, and with protocols for managing polluting elements, has a long history in Japan and sheds light on less well understood facets of the waste issues explored in this volume.

Pollution Avoidance, Past and Present

Traditionally, Japanese community life was something of a minefield of prohibitions and taboos, and yet Japanese armed themselves with a range of purification rites and other tools that gave them a measure of control over the dangers surrounding them. Sacred areas, which were the most pure, included Shinto shrine compounds, ritual enclosures, and wells. Special powerful features of the landscape such as springs, unusual stones, or old trees gave off energy and were identified and maintained through strict codes of practice. Cemeteries, blighted fields, taboo houses, and sites of death or misfortune all radiated pollution that was mediated through careful ritual action and flagged with an array of terms whose number and prominence hinted at their importance (see Namihira 1977 for an extensive listing). Many of these ritual terms are no longer in common usage, but in the erudite lexicon of folklorists, *kegare* indexes the polluted and *hare* connotes the pure, with *ke* designating the mundane realm of the everyday that is neither *kegare* nor *hare*. The sources of greatest impurity were death and blood; thus family deaths, funerals, cemeteries, and so on attracted a great deal of ritual activity not only to honor the dead but to safeguard the living. Similarly, childbearing or menstruating women, and those having miscarriages, were handled as sources of danger and pollution, often kept sequestered to avoid bringing misfortune to social activities such as hunts or other risky excursions (Namihira 1977, 1984, 1985).[7] Along the same lines, greater care and superstition accrued to dangerous or important activities such as seafaring in coastal villages, so that fishermen or whalers, for example, would perform rites before setting off on expeditions (Kalland 1982, 1995; see also Kalland and Moeran 1992; Namihira 1977, 1984; Parker 1983). Folklorists such as Miyata Noboru (1996, 1977), Sakurai Tokutarō (1970, 1988), and Namihira Emiko disagree somewhat on the precise contours of pollution beliefs. Namihira (1977, 1984), for instance, argues persuasively that *hare, ke,* and *kegare* constituted a continuum, with the mundane, everyday sphere of *ke* mediating between the "special" states of purity and pollution. But all agree that concern over *kegare* ritual pollution was a vital feature of village social life, comprising a category of threat that structured much of Japanese social interaction.

There is less consensus, however, on whether *kegare* beliefs remain so central to contemporary urban communities, where most of Japan's population lives. While scholars—particularly folklorists—generally concede that pollution avoidance is a key to understanding the poetry of a lost village social order that no longer exists in its previous form, Namihira and, more recently, Tom Gill (2005) aver that *kegare* persists as a relatively unacknowledged, but still significant, cultural force shaping activities and exclusion in postwar and contemporary settings. Namihira, helped at times by a team of Japanese scholars, carried out an ethnographic study (Namihira 1977; see also Namihira 1984, 1985) of three villages for several years, intermittently between 1963 to 1974, in order to map the variety and, she argues, the "structure" of Japanese folk beliefs rooted in conceptions of the polluted and the pure. The fieldsites varied—a fishing village on a small island off Kyushu, a community nestled in the mountains of northern Kyushu, and a coastal village on the island of Shikoku. All shared remoteness from metropolitan urban Japan and an abiding concern for sacred spaces and management of defilement, though with historical and regional variations. Namihira speculates on the extent to which such cosmological beliefs endure when transported from superstitious village community life to Japan's massive cities but clearly shows that *kegare* consciousness is not confined to the hazy precincts of Japan's long-ago village past.

Gill extends Namihira's analysis by tracing vestiges of "*kegare* avoidance" (Gill 2005) that circulate throughout contemporary urban Japanese society and provide a vocabulary and a set of protocols to identify and deal with pollution. He musters a range of evidence to suggest that Japanese remain preoccupied with, and wary of, *kegare* in varied contemporary settings that include *doyagai* day-laborer enclaves, leper colonies, foreigner hostels, *burakumin* settlements, and many other social contexts. Gill's discussion is less decisive regarding the question of whether such avoidance constitutes heart-pounding fear, mild disquiet, or mere pursuit of traditional practice devoid of revulsion, but these are points he will presumably clarify in later work. Gill's research is intriguing and provocative, but at this stage I remain unconvinced that, using the term *kegare*, one can connect, or even seem to connect, the visceral superstitions of Japanese village life of yore with present-day caution around various present-day "outcasts," with contemporary avoidance of *buraku* or *zainichi* Korean intermarriage and with low property values near *doyagai* or other "dangerous" zones.[8]

I believe it is more important to identify and analyze the different types of untouchability that have taken shape in contemporary times, an analysis best unshackled from attempts to create a metaconcept to explain much of Japanese society. That *kegare* have influenced the reckoning of high- vs. low-status elements and the boundaries between the pure and the polluted is clear, and some of these reckonings endure in the present day in atrophied, even "disenchanted" forms. But that the origins of some contemporary discriminatory practices can be traced back to *kegare* thought and practice does not necessarily translate into a contemporary social milieu dominated by this ontology of defilement. As was discernible among my informants, what might be called contemporary "*kegare* avoidance" preoccupied some Japanese far more than others, to say the least.) While *kegare* theory may not, then, be the skeleton key that opens up most or all Japanese prejudices and superstitions to analytical view, it does give anthropologists a reference point for understanding pollution and its transformation in present-day, relatively less superstitious urban settings such as Tokyo.

To be sure, ritual practices of purification and ideas of ritual pollution continue to circulate throughout present-day Japan in attenuated form, influencing the contours of the wastescape there. The spectacle of sumo wrestlers tossing purifying salt *(kiyome no shio)* into the ring before a bout is perhaps the most visible example, though pollution prohibitions also persist in a less publicized form, with sumo authorities forbidding women—deemed impure through menstruation and otherwise—to enter the wrestling ring lest they defile the sacred Shinto confines.[9] This practice, which goes by the term *nyonin-kinsei*, is still enforced in certain shrine compounds in Japan. Of course beliefs surrounding the polluting nature of women, particularly when menstruating, can be found in a wide range of societies, though most are preindustrialized.[10]

Shrines with more equitable rules of entry for women are nevertheless replete with thresholds and other accoutrements intended to manage pollution. *Mitarashi,* or fonts of purifying water, greet visitors to shrine compounds, there for ablutions before approach to the main shrine buildings; some of these waters have reputed healthful properties and a legendary provenance of miraculous cures for ailing communicants who drank or washed with them. Sacred precincts remain marked by *shimenawa*-roped enclosures and maintained through strict ritual

FIGURE 11   A local Horiuchi shrine warning against illegal dumping in the traditional pit for incinerating ritual implements.

practice. In Horiuchi's Kaeru-no-Jinja shrine compound, for example, rituals throughout the year would bring *shimenawa* enclosures for everything from designating temporary spaces for seasonal rites to surrounding the pit where ritual objects were incinerated after use. Ironically, the pit was also vulnerable to illegal dumping of household waste, which led the cleric to post signs there in 2009 urging proper conduct (see figure 11). These short-term sacred enclosures would be constructed alongside perennial sites of bound commemoration and protection, such as ancient trees or shrine portals.

As revealing as these examples of ritual or spiritual concern with pollution are, more important, I believe, are the ways in which ideas of purity and defilement radiated throughout daily life in less ritual-centric or anachronistic settings.

Domestic Disturbances

In my fieldsites, pollution avoidance was perhaps most visible in the home. No matter what the architectural or cadastral contours, the boundaries of the home (indeed, even the boundaries of rooms) resembled something akin to the trenches in terms of everyday Japanese negotiation of purity and pollution.

Hygiene and prophylactic health practices in Japan are particularly rich subjects of anthropological analysis because they involve highly relational forms of engagement that are, nevertheless, not usually aggrandized as "uniquely" Japanese. This allows us to get closer to an appraisal of Japanese engagement that is stripped of the nativism and romanticism that often accompany descriptions of Japanese society by Japanese and foreigners alike. To take a closer look at hygiene and health in the home, I return to a point made by Douglas, who writes:

> There is no such thing as absolute dirt: it exists in the eye of the beholder.... In chasing dirt, in papering, decorating, tidying *we are not governed by anxiety to escape disease,* but are positively re-ordering our environment, making it conform to an idea. (Douglas 1966, 2 [emphasis mine])

While her dismissal of health fears as a root cause may suit her otherwise brilliant argument, this distinction is less well supported by domestic practice, notably in Japan. To take one example, parents in many parts of the world feel that a sanitized home offers their children protection against illness, regardless of what methods they use to achieve this or what the results may be. Dismissing this dimension of analysis leaves important and revealing sociocultural attitudes out of the equation. My discussions with parents in Horiuchi made it clear that while order—as construed by Japanese—was a central concern, many residents' conceptions of domestic order very much included a space made free, as much as possible, of disease-carrying germs that might threaten themselves and their families. Family welfare could be taken to extremes: some Japanese parents kept their children inside during culturally determined "dangerous" times of the year, such as the turn of the seasons or on particularly cold days. And while local parents tended to state that their neighborhood was preferable to other, less familiar areas, these parents also saw relative threats close to home—that is, in the community surrounding them, even right outside their doors.

Residents in Horiuchi and in community Japan generally viewed the home as a bastion of sanitized order against the perils beyond its confines. To put it bluntly, "inside" in Japan is typically associated with cleanliness and social warmth, while "outside" implies social distance and has associations of impurity and possible disease.[11] This, it should by now be clear, went far beyond the practice of removing footwear before entering, though concerns with footwear reflect this ontology of defilement. Such precautionary measures stemmed in large part from a desire to limit exposure of the interior of the home to *baikin,* or germs, a concept closely associated with the dirt of the outside sphere, as Ohnuki-Tierney (1984) develops in great detail. Numerous informants' families created something of a decontamination lock near the entrance of their home, sanitizing themselves in a disciplined manner when returning inside the home to annihilate germs, particularly in winter when communicable diseases were deemed rampant. For example, Koda-san, a Horiuchi-based informant in her forties, explained it almost like scrubbing down for an operation: "My father and uncle are doctors, so when we were kids we all washed our hands and gargled right after we came back [into the home]. That way, we knew it'd be safe *(anshin)."* While this clinical approach was typical of Japanese with some medical or technoscientific background, it was by no means exclusive to such groups. In tandem with the widespread washing of hands and gargling, Japanese informants also used white gauze face masks when out and about in potentially polluted spheres. They do so not only to prevent their own germs from spreading when ill but sometimes as a prophylactic measure to avoid germs on crowded commuter trains, at work, or in public spaces. The proprietor of a small Horiuchi bar that specialized in *yakitori* grilled chicken was one of these, saying he would usually don a mask while working once the colder months rolled around: "Of course, if I have something myself, I don't want to pass it on to my customers, but [the mask] also comes in handy—I mean, no one's gonna pay me sick leave if I get saddled with a cold at work, right?"[12] Many of these concerns seem related to the reflex-like Japanese stance against outside threats broadly defined. Compared with those in other societies where I have lived long-term, such as the United States, England, and France, some residents in Horiuchi—but by no means all—seemed unusually well informed (and voluble) about the geographical source of any given strain of influenza making the rounds. One year the Melanesian/Australian origins of the flu were a topic of conversation. Another year it was a Russian strain and yet another, a strain from Hong Kong. In a nation with such structural xenophobia as Japan (e.g., Tsuda 2003), it is difficult to discount such attention to the foreign origins of a dangerous flu, particularly against the backdrop of racial constructions of purity there.

Items and persons could be domesticated and purified through use of traditional decontaminants, like prolonged sunlight and sprinkled salt,[13] or through antiseptic products or prophylactic procedures either upon entry or afterwards. With forms of waste that went beyond the pale, such as dog feces, residents posted numerous signs urging dog owners to collect their dog's droppings to keep the community "pure" (*seiketsu*) (see figure 12). Yet the boundaries between the polluted and the pure were not simply congruent with the exterior of the home. Even within the sanctum of the home, there existed contaminated spaces that required proper handling. Zones of purity and impurity within the home were sharply demarcated, with toilet rooms, the kitchen, and the verandah less "pure" than *tatami* mat spaces or where family members slept. Almost all Japanese toilet rooms offer a dedicated pair of slippers to wear upon entering. This allows undesirable germs on the floor to remain within the toilet area rather than spreading to the rest of the home. Furthermore, nearly all my informants' homes, even tiny studios, had separate outdoor sandals or slippers on open-air balconies that the

FIGURE 12   A common Horiuchi sign urging dog owners to keep the community "pure" (*seiketsu*) by collecting their dogs' droppings and getting dogs their required vaccinations. Vaccination notices are posted outside the homes of pet owners to monitor compliance.

inhabitants would wear when hanging laundry or sneaking a smoke. Since many Japanese like to wear yet another pair of indoor slippers in the home (particularly during colder months) and usually remove these before stepping into rooms with *tatami* mats, walking about a Japanese home can become a laborious balancing act of footwear changes, though one to which Japanese residents become fully accustomed.[14] This everyday choreography gives testimony to the perceived importance of keeping impurity in its proper "place."

The very construction and design of homes is undertaken with an eye toward purity and avoidance. Building sites in Horiuchi, Izawa, and elsewhere are routinely consecrated through ritual practice (incantations, sake-pouring, and so on) with a *jichinsai* groundbreaking ceremony, a purifying rite conducted by either a Shinto cleric or a Buddhist priest. Whether the rite was Shinto or Buddhist usually had little bearing on the "beliefs"[15] of the commissioning homebuilder, but the Buddhist ceremony tended to be a little speedier, which could become a consideration.[16] Attention to purification and the warding off of danger continue throughout the life of a construction project. Construction workers can be a superstitious lot, by and large, perhaps partly due to the relatively dangerous nature of their work; the more dangerous the work, the more superstitious those prone to superstition tend to be. My good friend Shin-chan, a fifty-something, gregarious, life-of-the-party reveler in local bars by night, was an experienced builder by day who was very strict with his subordinates about site safety. Though not concerned with superstition himself, he nevertheless explained to me the dangers of a building site without the benefit of proper ritual protection: "When you're working on a [high, sloping] roof with a couple of other guys, the last thing you want is for them to be thinking about anything else other than exactly what they're doing." And on the client side, Japanese property owners and developers usually decided not to take their chances.

With all this at least partly purity-focused social behavior, it was curious that, on the edge of Horiuchi, a long-standing *burakumin* block coexisted relatively uncontroversially with the rest of the area. Called Oden Peddlers' Alley (*odenya-san no yokochō*) by some, the block was a narrow lane lined with more or less decrepit buildings occupied, effectively, by squatters. *Oden*, a kind of stewed cuisine sometimes sold from portable stalls, was hawked at night in entertainment districts like Shinjuku in the early postwar period, and the Japanese who peddled it pushed their carts in and out from far away and stored their carts in places like Oden Peddlers' Alley. In time this entrepreneurial practice fell out of fashion. But also, according to my informants, after the war Tokyo faced an acute housing shortage and, at one time, told some homeless Japanese to take refuge on the land, many of whom worked hard peddling *oden*, ergo the name. As Tokyo's crisis passed, the land remained in a legal limbo. Fukushima-san, a community leader and grey eminence in Horiuchi, confided, "You know, if you go into the ward office and look at the official [cadastral] maps, they have every single household marked [there]. But [Oden Peddlers' Alley] isn't marked. There's nothing there, like it doesn't exist!" (In fact, this was not the case when I checked in 2009. But it is revealing, nonetheless, that some members of the community believed this to be so.) The original peddlers, their blood relations, and others squatted on the land, thus avoiding certain taxes but at the same time denied nonessential public services. There held an uneasy truce; the existing squatters could remain, according to informants in Horiuchi, as long as they lived in the (deteriorating) existing structures and maintained the status quo. If they rebuilt, they would lose claim to their abode.

Some locals avoided the alley as "dangerous"—they suspected a heightened criminal threat there, which I never found substantiated in the least. Likewise, some female friends refused to park their cars anywhere near the area. Kumi-san explained, "[When driving nearby], we're really really careful. The reason is, they'll want to break in to your car (... *watashi-tachi ha kono toki ha sugoi ki wo tsukeru. Dōshite ka to iu to, ake ni mitai*). It's never happened to me, though, so I don't know for sure.... For ages, everyone's said, 'Be *really* careful there.'" She went on to say, "I don't know any of them, but I'm pretty sure they're *burakumin*. Just look at the homes there. They're really dirty, there's not much pride there *(datte, sugoi kitanai shi, hokori mo sukunai shi)*." Other residents sometimes avoided walking down the block at night on their way from the station or weaving home from the bars. But the people who lived there comprised a tight and upstanding community, bound partly by their campaign against Tokyo to remain on the contested land, and to me they seemed not the least bit threatening.[17]

The relatively dilapidated warren was located just next to a small commercial district called Hokkaidō Town (to drum up business), and the locals had largely adapted to the putative *buraku* presence, which indicated that contemporary Japanese attitudes toward *burakumin* (real or imagined) are not as virulent as sometimes assumed. One mitigating factor was the ambiguity of their status. Shin-chan, a close friend, explained, "No one really knows if they're *burakumin* or not *buraku-*

min. People who worked [peddling *oden*] might likely have been *bura-kumin,* but then they could have passed their residence on to someone else." Making a gesture for a scar down the cheek, Shin-chan added that these people were nothing like those other, more dangerous *burakumin,* referring to the oft-unspoken belief that the ranks of the Japanese mafia are filled with "untouchables" of various kinds. Some inhabitants of the alley kept to themselves—a few locals speculated that they probably could not afford the prices in the local bars—and there was sometimes almost no sign of life along the long block. But the condition of their homes and other visible elements still flagged their difference. Shin-chan, who felt no knee-jerk ill will of any kind toward outcasts, walked down the alley with me one afternoon. Gesturing toward an improvised parking space with an old work truck surrounded by scrap and other (apparently scavenged) building materials, he muttered, "Look at that. The guy takes up half the road with his junk. That's just selfish. . . . You couldn't get away with that elsewhere." For her part, Kumi-san confided, "Of course, the people living nearby are scared *(kowai).* . . . I don't think [the outcasts] can use the local bathhouse *(sentō),* for example. I'm not sure, though."

It is clear from the above that Japanese I interviewed still "saw" (and felt) forms of pollution, to a certain extent at least—including social pollution—in the world around them. Furthermore, some practices invested them with some control over their environment. Tossing salt, pouring sake, uttering incantations, gargling, using heavy-duty cleansers—indeed purification, broadly defined—these were tools to manage threats and uncertainty. To the extent that dirt, disease, and misfortune were a part of everyday life, my informants felt they could be kept at bay, to a degree, through careful effort. In Horiuchi at least, this was not heart-pounding terror or superstitious dread of *kegare* pollution. Yet against such a backcloth, the toxic waste and pollution scares of Izawa show clearly how much anxiety could materialize in a liminal zone in which familiar social boundaries had collapsed and formerly "normal" residents became, almost overnight, "untouchables" subject to aversion, aggressive ostracism, and exclusion.

Disorder(s)

I have endeavored to demonstrate the suggestive overlapping of ideas of ritual pollution *(kegare),* "social pollution" (categories of untouchability and resulting exclusion), and social reckonings of hygiene, filth, and illness. It is no mere linguistic accident that environmental pollution also plays a role in notions of purity and defilement. The word "pollution" in English derives from the Latin term *polluere*, "to make ceremonially impure, to violate, degrade, to make foul or dirty, to soil, stain" (OED 1989), suggesting that broad semantic overlap between different notions of impurity has existed not just in relatively well-known Asian cases such as on the Indian subcontinent, but in Europe and elsewhere. Increased conspicuousness of toxic waste and other poisons in contemporary social life—either in numerous communities in material form or in their mediaspheres, discourse, and consciousness—has ushered in widespread aversion to environmental hazards. What is rather unusual in the Japanese case, however, is the extent to which environmental pollution can convert to social pollution. Victims, often through no fault of their own, can suddenly become ostracized as outcasts after spending most of their lives as accepted members of the mainstream. Not only does this process mandate a thorough scholarly assessment of ritual pollution and social pollution, as I have attempted thusfar in this chapter, but it also begs a deep and wide-ranging inquiry into the social valences of waste in a complex society like Japan, the latter objective serving as a central focus of this monograph as a whole. Unfortunately for some of my informants, one unpleasant dimension of life in Tokyo's wastescape is the "toxic," as it were, social pollution that can result from the vagaries of proximity to waste facilities that, in urban Japan, all too commonly exist in densely settled areas.

Abrupt social exclusion of otherwise "normal," mainstream Japanese can take different forms. Salarymen who divorce can find themselves informally designated *batsu-ichi* by members of their social circle or company hierarchy, a term that can be loosely translated as "one strike against." This can influence their status, relations, and career trajectory.[18] Similarly, single mothers fit differently in rigidly male-oriented Japan because they deviate from traditional family models; victims of structural marginalization, these women can also face acute discrimination at work and in finding housing, for example (Perry 1975; Ezawa [personal communication]; see also Ezawa 2006; cf. Imamura 1987). The taint can hover like a cloud over the children involved: applicants from nonstandard or "broken homes" have faced disadvantages in applying to schools and even universities in Japan, which is one reason why a few informants spoke of Japanese couples waiting until their kids were past that age before filing for divorce. Exposure to the foreign can bring its own hazards. Japanese who remain "too long" overseas run the risk of being treated as no longer wholly "Japanese" on their return. This can be particularly the case for women who "learn" to be less subservient or compliant around Japanese men, or who are even suspected of being so, as happened (among others) to a female informant of mine in her late twenties who worked as an advertising executive for five years in Florida.

Illness is a common trigger for social exclusion, but rather striking in Japan is the notion that noncommunicable disease (or its perceived presence) can contaminate social relationships or interactions with strangers. Gastrointestinal illness, for example, has carried a stigma: in one publicized case, when thousands of children came down with *E. coli* food poisoning in a July 1996 outbreak in Sakai, near Osaka, the affected children, once they had recovered, found themselves shunned when they returned to school, deemed "contaminated" long after the condition had passed. For one unlucky family, fellow residents of their apartment building reportedly forced the mother to scrub the stairs and wipe the railing every time her daughter walked on them to "protect" other children in the building, despite the fact that this is far from a standard preventative measure for such gastrointestinal agents (*Japan Times*, 12 November 2000). Even victims of terrorism have found themselves subject to similar processes of exclusion. Novelist Murakami Haruki, in his chilling nonfictional oral history of the victims of the sarin nerve gas attack on the Tokyo subway in 1995, discovered that those who survived the attack—but with health complications—could find it extremely hard to assimilate back into their groups at work and in communities, partly due to the ostracism of unsympathetic former friends and peers (Murakami 2000). Legal transgression is also linked to this phenomenon. Not only can criminals be ostracized once their crimes become known to the community, but so can their families.[19] Though this stigmatization can occur in other societies as well, in Japan it seems linked to the broader processes of exclusion I have been describing.

The effects of environmental illness in community Japan bore similarities to these examples, but with one important difference. This was the degree to which victims tended to be localized and share a sense of grievance, offering great potential influence as a protest group (e.g., George 2001; Broadbent 1998; Huddle and Reich 1987). Whereas survivors of the sarin gas attacks were dispersed throughout Tokyo, even Japan generally, and endured any exclusion largely in isolation, neighboring victims of toxic pollution could more easily forge networks of support and solidarity to combat their new outsider status. Such was the case with victims of Azuma Disease in Izawa.

Izawa's afflicted saw their world change dramatically, in some cases literally overnight. Several protesters used the trope of a "bad dream" to describe the outlandish, topsy-turvy, Alice-in-Wonderland quality to the direction their lives turned. Kamida-san explained, "It was like living in a nightmare *(akumu)*. Everything looked the same in my life, but it was all different. There was nothing I could do and I couldn't wake up....I looked at my body sometimes and it was like it was someone else's." Their bodies could seem strange, indeed, producing lumps, boils, mucus, pus, and rashes. Substances that appeared out of pre-existing orifices might be green, yellow, brown, foamy, clumped, or hard. The odors they perceived their bodies giving off were unfamiliar and frequently abominable even to themselves. Around them they detected various kinds of stench that they had never previously associated with their homes and communities. Mingling with their pain, nausea, and disability was a strong sense of abhorrence at the filth they found in their lives, both the toxic waste coming from the waste facility and the wretched, unclean byproducts of their bodies' fight against illness.

The mysterious toxicity of the pollution circulating through Izawa made those afflicted with Azuma Disease feel violated—sufferers began to conceive of their environs as contaminated, even hostile. If the interior of the home can symbolize a relatively unpolluted realm for many Japanese, then the prospect of airborne toxic pollution in the community was a deeply unnerving one. Even the "purest" part of the home, which might be the formal *tatami* room often reserved for guests or simply the baby's crib, was likely to be just as tainted as anywhere else. This led to a degree of despair voiced by otherwise stoical victims of Azuma Disease. As Awaki-san stated wearily at one stage, "You know, if the poisons are everywhere, you can move all you want but there's nowhere to go....I don't know what else to do, really."

The impact of this pollutant transformation was concussive. What once seemed familiar and reassuring became hooded, oppressive, a source of unease. No one seemed to know the extent of the danger or how to avoid it. Residents suffering from toxic exposure in Izawa complained that they had to keep their windows tightly shut at all times, this in a society where open windows traditionally provide a cherished alternative to air-conditioning, particularly at night in summer months when the air cools down—though this domestic lockdown did little, if anything, to alleviate either symptoms or anxieties. The afflicted were also quick to point out that they felt unable even to hang their laundry or bedding, thereby losing the benefit of "purifying" sunlight and fresh breezes. With toxins apparently swirling around in the air and omnipresent in the community, many residents were unwilling to expose their homes and their bodies to the toxins, so instead they hung their laundry indoors to dry *(heya-boshi)* or used dryers. This feeling deepened when residents noticed the traces that "dirty" air left on their balconies and laundry frames. One man in his twenties with relatively mild symptoms complained: "[These days] I always get diarrhea and a throat that feels out of sorts *(fukaikan)*. [The laundry frame] quickly collects soot....I work at home all day long, so I'm really worried." A woman also in her twenties in the same area ceased air-drying on her balcony altogether: "Ever since I stopped hanging my laundry outside to dry, my itching condition disappeared." Those most sensitive, like Tsubō-san, had to discard their old clothes, bedding, furniture—everything, really—to avoid toxic reactions from trace amounts left on their possessions. But even when taking extreme measures, few of those afflicted with Azuma Disease felt free of contamination while in Izawa. While attempting to seal off the interior of their homes appeared to help somewhat in certain cases, sufferers reported how, at times, the space of the home became so foul that they despaired of ever returning to the "normalcy" they remembered before the waste transfer facility commenced operations. Some complained, furthermore, of smelly, filthy water pouring from the tap and sharp reactions while bathing. Others found their "allergies" were far worse, even while sequestered indoors, and found that they were falling ill with far greater frequency.

The enveloping, all but inescapable, toxicity of the air that sufferers experienced led to a sharp transformation in the social topography of the community. Like some celestial black hole of immense force come to earth, the waste facility buried in the center of the community loomed large in the minds of those afflicted with Azuma Disease, pulling at their thoughts even at a distance. There were those residents who could no longer move save to shuffle around in their homes, better off than the bedridden but only just. They continued to feel (somewhat) human only through visits from friends and family, visits the sufferers tried to discourage lest loved ones fall ill as well. But ambulant sufferers faced difficulties of their own. If afflicted residents took a walk, they might now choose a far-flung route to avoid passing near the waste facility. They would even plan their itinerary literally with regard to which way the wind was blowing that day. Early on, those with acute cases like Kōga-san, a male retiree in the area, might wait to go outside until they thought the wind had changed direction so as to decrease their exposure to what they termed the "smoke" *(kemuri)* coming from the facility. Once he explained, "It's better now. This morning [the pollution] was pretty bad, though....At night, too, we have to be careful. It's pretty strong at night a lot, so we always check the air—it's like they do more at night so no one will notice....We're lucky on this side, though. The other side of the [waste facility] gets more with the wind than we do." Some sufferers pointed to seasonal variations in patterns of exposure, one meteorological contribution to the upsurge in local brainstorming, even conspiracy-theorizing, as residents struggled to interpret the problem. A man in his twenties who had lived in Izawa for three years claimed, "My [skin problems] have gotten dramatically worse *(atopī no akka ha ichijiru-shiku)*...especially in winter [with] the north wind....I'm going to the hospital all the time like when I was a kid....My symptoms [change] a lot with regard to the direction of the wind." Given these hostile environs, those with cars might drive short distances within the community rather than walk in the open air or choose to patronize shops and businesses far from the immediate community. One couple closed down the small business they ran near the waste facility, partly because running it entailed spending long hours, daily, within the perceived toxic zone. If, as Lynch suggests, the spatial experience of living in a community can be understood through the landmarks by which people choose to orient their (practical and mnemonic) navigations (Lynch 1960), then the waste facility itself became a negative landmark, a site to be avoided, and yet one that loomed in the mind as an undesirable specter of insalubrity. And as residents' iterative ambits throughout the community shifted with regard to the field of "reverse-gravity" their aversion to the facility created, it became increasingly difficult for the afflicted to recognize the intimate contours of their community under the pall of toxic anxiety they perceived around them.

The resulting social repercussions were severe and created complex knock-on effects that hit different community factions in divergent ways on what had become a politically fragmented social terrain. Those battling Azuma Disease found it very difficult to separate their feelings of ill health from their attitudes toward Izawa itself. Echoing Watsuji's nativist conceit of *aidagara* ("betweenness") (Watsuji 1975), which he believed distinguished social engagement in the Japanese milieu, Japanese informants described Tokyo as comprised of sites of intensive relationality. Between the familiarity of a person's home and community and the familiarity of their place of work could be an indeterminate, acontextualized liminal zone of "colourless indifference" (Berque

1997b, 90) where a lack of relations led to a lack of emotional engagement. Along this theme, Lock (1980, 255) writes, "The average Japanese is interested in the problem of pollution, for example, not in its global aspects, but only insofar as it affects one of the groups to which he or she belongs." The other side of this coin, however, is that when one's community is afflicted by pollution, the relational mapping and affective ties of the area are changed utterly. Conditions suffered by residents in Izawa demonstrate how easily social boundaries can be undermined by pollutants and illness, leading to unexpected remappings. Indeed, informants' responses suggested that toxic zones become bizarrely liminal, an unsettling mix of features that could leave those afflicted feeling as though the groundedness and intimacy their communities once offered were forever beyond their grasp.

Sufferers of Azuma Disease found out all too quickly, furthermore, that they no longer enjoyed the same status they once had. Close friends and relatives remained, generally speaking, reliable sources of support throughout the ordeal. But neighbors, fellow members of formal groups to which they might have belonged, acquaintances they might have encountered regularly—many of these people turned against them before long.

The reasons for this abrupt shift appear to have been varied but overlapping. To some members of the community, victims of Azuma Disease were ailing, contaminated, and therefore a threat. The ill-defined nature of their affliction early on made their condition all the more worrying. While someone coming down with a case of the flu was handled with well-established protocols—perhaps one avoided contact for a number of days, washed one's hands, gargled, wore a mask if necessary—the rules of engagement, as it were, for Azuma Disease remained uncertain. One informant in her forties, who was not a member of the GEAD protest group and was careful to remain disassociated (and adamant about anonymity), put it like this: "Back then, we weren't sure what was happening.... [one neighbor] had fallen ill and then about a week later [another neighbor] came down with similar symptoms, then another, and so on. We didn't know if an illness was making the rounds or something else. Then they started criticizing [the waste facility]." A female GEAD supporter in her late fifties, Yamaguchi-san, could understand this confusion from the perspective of other members of the community. She commented, "No one knew the cause. And once a few of us had similar symptoms, people with families and children and so on didn't know what to do. They didn't want to...[using her hands, she made a gesture which conveyed *pass it on*]." Mr. Awaki, who was president of a small technology company and married to a central Azuma Disease casualty, described the reaction of the community in terms of Japan's history of avoidance and ostracism *(mura hachibu):*

> Back when people lived in villages, say, in the mountains...there were contagious disease epidemics *(densenbyō)*...like tuberculosis *(kekkaku)*, for instance....If someone got sick, the family wouldn't say, "We've got somebody sick here!"—if at all possible, they wouldn't say that. Because if they did, not just that one person, but everyone in that family would be driven out of the area completely *(sono hito dake de ha nakute, kazoku minna ne sono chīki irarenai yō ni oidasarechau)*. Well, what's called *mura hachibu*, getting driven from the village....just like that, here with Azuma Disease I don't know if they'd call it a contagious disease *(hayari-yamai)* per se here, but there's a mysterious *(etai wo irenai)* illness....So if some family member thinks, "Something's wrong with me—this might be Azuma Disease," they can't tell their family *(kazoku ni ne, uchiakerarenai)*, right? And even if everyone in a family were thinking, "Geez, I feel awful," they might not say that to the people around them.

His account, while clearly speculative to a degree, rang true with other elements of social life in Japan I have experienced, where Japanese can often maintain an impassive united front for the sake of the group despite individual disagreements or misgivings. In addition, he backed up his analysis with anecdotes about the secretive donations of money to the cause by members of the community unwilling to support the GEAD publicly. A common example, according to the Awaki family, was a woman whose husband was anti-GEAD but who still wanted to help. And Mr. Awaki, most importantly, clarified the murky political dynamics of Izawa. Power in the area was held by just a small number of families, controlled by a similarly small number of heads of families, and he argued that this influenced greatly the character of public discourse in the community. With the structure of Japanese descent and ownership based on main households with linked branch households *(honke* and *bunke,* respectively), financial and other affairs of extended families could often be dictated by the main household. He explained, "If you're the new wife *(yome)* in a branch household and you come down ill [with Azuma Disease], there's no way you can go public with it.... You might be able to tell your husband, but that's it." One issue was that the offending couple might be stripped of the right to use the land and house they might hold, for example, at the pleasure of more powerful relatives opposed to the protest. This, protesters believed, was the primary source of Izawa's rather monolithic opposition to the GEAD, as well as the anonymous donations—with apologetic notes attached—that some received in their mailboxes.

In fact, Mr. Awaki believed—repeating a commonly held point among the protesters and their supporters—that the whole issue locally was real estate values. "Some of these families have been here for hundreds of years.... They held agricultural land, and then they made a lot of money when Tokyo expanded.... So a lot of these people got rich quick... And if it's proven that Azuma Disease exists, their land goes down in value, people don't want to live in their apartment buildings, rents go down, and so on.... And they can't let that happen." Mr. Awaki used the image of the traditional "see-no-evil, hear-no-evil, speak-no-evil" monkeys *(mizaru, kikazaru, iwazaru)* to mock such heartless, avaricious types (see figure 13).[20] These nouveaux-riches *(narikin)* members of the community were, according to the Awaki family and others, once

FIGURE 13   One GEAD supporter compared Azuma Disease naysayers to the famous three monkeys *(sanzaru)* at Tōshōgū Shrine in Nikkō.

very active in the community, but their involvement had atrophied with their wealth and the changes in the area. They remained highly conservative politically—being, generally speaking, staunch supporters of the ruling conservative regime in Japan up until that time, the Liberal Democratic Party—and rigid socially, intent on preserving their influence in the community. So anyone who breathed a word of support for the GEAD or who complained of toxic symptoms, according to informants in the community, stood a good chance of being ostracized. *Mura hachibu* in its traditional form, with the village driving the undesirables out into social oblivion, is largely impossible in contemporary Japan, but in Izawa some members of the community did their best to make life very cold for those who spoke out against Azuma Disease, the waste facility, and the duplicity of those who refused to acknowledge the environmental illness plaguing the community. The long-term presence of this social frost was made patently clear to me one day. We were sitting in the Awakis' living room drinking Japanese tea with the windows shut to the spring air outside. Pointing from her place on the couch, within a few hundred yards of the waste facility, Awaki-san gestured down the street to identify the families who had essentially amputated the Awaki household from their lives. One was just a few houses down, then another across the street, then another on the other side, and another there across the back garden. The Awakis' house was surrounded by opponents who had made clear their desire to see the back of them. Then Awaki-san gestured down a side street, invisible from her spot on the couch but discernible mnemonically, intoning landmarks and distances to describe where Tsubō-san and Kamida-san and some others used to live. The former two had been forced by their symptoms to flee the area altogether. Frequently, during her descriptions of the toxic pollution and local social politics, Awaki-san made frequent reference to the Izawa that once was, before the filthy air came and turned everything upside-down.

The afflicted and their supporters were in a destabilized, ambivalent position—pariahs themselves, they nevertheless regarded their polluted surroundings with the same horror that an "ordinary" Japanese might view a postwar leper colony or a *burakumin* settlement (in the extremely unlikely event that they would ever encounter one of these). Cultural biases inculcated in them by their many years in "mainstream" Japanese society could produce intriguing contradictions. For example, it is accepted in some Japanese circles, including in the opinion of a number of native Japanese social scientists, that *burakumin* and other low-status workers are actively involved in the waste industry. Contemporary hard-luck or outcast day laborers also work on waste-related projects in addition to more common construction projects, and low-status day laborers cleaned and maintained the inside of nuclear reactors in radioactive conditions (Horie 1984; Tanaka 1988; cf. Gill 2000), high on the list of dreaded contemporary polluting waste-related tasks.[21] Apocryphally, owners and operators of incinerators and other waste-related private facilities can be *burakumin* as well, though this remains unsubstantiated—indeed, with Japanese discrimination against *burakumin* not yet a thing of the past, reliable figures on the backgrounds of "outcast" employees in almost any sector would not only verge on being unethical but would be nearly impossible to compile. Yet the fact that this notion circulates in Japanese discourse is by itself revealing. In a discussion with Awaki-san and Kamida-san about employees of the waste facility, I probed as to whether any employees might be *burakumin* and whether this would make any difference to them. Awaki-san's face, usually serene even when being critical, immediately darkened: "Yes they're *burakumin!* They're all *burakumin* over there!" Her frustration and disgust were palpable, and the term *burakumin* was a clear epithet. Yet, during another interview about nine months later, when *burakumin* came up again in conversation, she stated, "That's what I am, a *burakumin.* I'm the *burakumin* here. That's exactly what I am," turning the expletive into a social badge of courage that reflected her stigmatized position in the community.

Whatever one's opinion of the purity or defilement of Izawa's precincts, in social terms the community had been riven asunder, if not always visibly so. A comment from Fujimura-san, a woman in her late fifties who was not a member of the GEAD protest group and who did not "feel" Azuma Disease, was representative of many who were touched by the virulent political controversy around them: "I've lived here for twenty-one years. Some things have changed a lot since [my family] moved here, but the area always seemed the same. Now things are different." Not least was the stigma of illness: once some residents began to feel the symptoms, they enjoyed sympathy from some but surprising coldness from others. Such hostility made journeying through the community something of a minefield for sufferers who were already hard put physically to make it through the days. As Kamida-san, a female GEAD stalwart, explained: "At first, I was really uncomfortable bumping into my neighbors. I'd do almost anything to avoid it.... Getting [the cold shoulder] from people who'd been close to me, that was [really difficult]. After a while, though, I relished it—I began fighting for all of us."

A process of inversion took place by virtue of which the afflicted in Izawa found their wretched pariah status converted into honor and prestige (see Stewart 1997). Those who "felt" Azuma Disease were imbued by their comrades with a kind of mystical power, hypersensitive and therefore special. Not only did their condition give them corresponding clout in the protest group, but it made them into something like minor celebrities when attending conferences or talks with environmental activists. When discussing their plight, some were deferred to in a way that generally they would not have been otherwise. Whatever social advantages—if they can be called advantages—that they gained, however, were double-edged. The tension between being excluded as a outsider and fighting the establishment as an outcast—this predicament of being perceived as polluting while perceiving pollution in one's surroundings—took a great toll on GEAD activists and their supporters. Members of the community like Tsubō-san, presented with contaminated surroundings, polluted homes and belongings, poisonous social relations, and a Tokyo government apparently more concerned about protecting the integrity of macroscale metropolitan waste strategy than the microscale of community environmental health, took the only decision they felt was left to them: they moved.[22] Of the formal members of the GEAD, after a few years, only one who actually "felt" Azuma Disease (Awaki-san) stayed put where she was. "Someone's got to stay!" she exclaimed. "I've got to make sure people remember here, right?" But she sometimes bemoaned the sacrificed good health, the memory loss and mental fogginess. Several others stayed relatively nearby, a bike ride, short bus ride, or quick train trip away. The rest, and other nonmembers, chose to relocate to new and sometimes remote communities with clean air and other treasured amenities, like easy access to gardens or forest. This ambivalent self-exile—a sort of contemporary variant on *mura hachibu* exclusion, with recalibrated power vectors—led to a strangely disembodied, often virtual protest. Tsubō-san chose to live in the suburbs of a nearby prefecture with clean air and open terrain but commuted regularly, bundles of files in hand, to ward and metropolitan meetings and environmental symposia. Some, like Kimura-san, continued to work in the community—still as a playground monitor and self-styled ecology instructor at a local primary school. As such he constituted a small but influential "live" daily presence in Izawa. These self-exiles, both relieved at their improved health and embittered by the political wrangling with recalcitrant local leaders and mendacious bureaucrats,[23] frequently looked back at their aborted time in Izawa with some nostalgia. But for the most part, and from a safe remove, they reflected on the immensely distressing memory of wrestling with a toxic threat that they are determined never to encounter again. And this was a memory that was held bodily. Tsubō-san's frail voice lowered when she stated, "I have no interest in going back.... Sometimes my body feels [like spasming] even to think of it."

Culturally and socially embedded notions of purity and pollution shape the topography of Tokyo's wastescape in subtle ways that influence conceptions of health, environment, and identity there. The next chapter carries forward this theme to show how ideas of purity and defilement circulate throughout historical and contemporary discourse on reproduction, national strength, gender roles, and contemporary sexual issues that are impacted by toxic waste in urban Japan. Toxics and their publicized role in attacking reproduction combined with sexual anxieties that influenced ideas of what is "natural" to make toxic waste a contributor to the gloom that gripped recessional, post-Bubble Japan.

CHAPTER 7

Growth, Sex, Fertility, and Decline

Present-day Tokyo contrasts markedly with the idealized constructs of traditional community life that circulate throughout Japanese social discourse and mass culture. While some communities are bound together with close-knit family ties and imbued with warmth, fertility, and a cooperative ethic—thereby approaching the nostalgic Japanese ideal—present-day Tokyo betrays tension between affective relations and the divisive pressures of contemporary life in the Japanese megalopolis. This chapter scrutinizes the slippage between these community ideals and the comparatively fragmented, insalubrious urban lives of many Tokyo residents, in particular how anxieties and resentments over fertility, sex, and family health are ignited by toxic waste issues.

Couples and families may, with justification, view fertility as their own intimate turf, so to speak. Yet states have made frequent incursions into this domestic territory. The present chapter traces Japan's preoccupations with national fertility and pronatalism from pre–World War II colonialist designs, which intensified during the war, into the high-speed growth period, which had its own particular demands on state priorities and state control. Japan's end-of-millennium convergence of relative economic decline with a much-publicized birth-rate decline, coupled with an aging population, has made fertility, procreation, and children's health highly charged political issues. Consequently, when media outlets and other sources disclosed how toxic waste attacks the human reproductive system and disrupts fetal development and child development—even genital development and sex birth ratios—they created widespread dread of such wastes, particularly dioxins and other endocrine-disrupting "environmental hormones" *(kankyō horumon)*. This has helped spur a series of government measures that shed revealing light on waste issues and social attitudes. Here, I scrutinize specific anxieties surrounding fertility, family, and anatomy in my fieldsites and throughout Tokyo's wastescape that sprung from actual or rumored toxic dangers suffered in proximity to waste facilities and elsewhere. I then examine how waste issues impinge on Japan's visions of its future. Any discussion of contemporary Japanese fertility and related concerns must address Japan's pronatalist and progrowth past, so the following sections endeavor first to contextualize properly the state-led preoccupation in Japan with both population growth and economic growth through to the postwar period.

Extended Family: The Growth of Japanese Empire

The Japanese obsession with growth has had wide-reaching socioenvironmental repercussions whose origins stretch back further than the recent past. The fateful speculative Bubble that began expanding about 1987 and burst in 1991 marked the headiest stage in a long trajectory of economic development reaching back to the early postwar period. Growth, or *seichō*, became both the mantra and a defining characteristic of Japanese postwar society—growth was, after several decades, virtually all that postwar Japanese knew,[1] and relations between Japanese citizens, the state, and corporations took place with the understanding that growth would be achieved, at almost whatever the cost. Yet the social and historical underpinnings of this drive went much deeper. Japan's headlong pursuit of growth can really be understood only within the context of Japan's maximalist Meiji-period and post-Meiji geopolitical ambitions; its embrace of the "Western," particularly European and New World industrial production and technological innovations, since the mid-nineteenth century; and its sometimes mimetic, sometimes adversarial posture toward the West for even longer. The affective rhetoric of the "family-state" that accompanied and, in effect, veiled more Machiavellian concerns reveals important elements of national-cultural orientation that continue to shape contemporary Japanese attitudes and their social and environmental impact.

Much of the early development of Japan's nineteenth-century industrial capabilities bore clear military undertones. Pursuit of modernization involved, among other things, sending bright emissaries to identify, record, and in some cases master the industrial, technological, and even social advances of other nations that would help Japan attain its goals. In addition to developing or upgrading Japan's pre-Meiji iron foundries and smelters; mechanized textile mills; coal, iron, and copper mines; and limited shipbuilding facilities, the government set about the monumental task of providing Japan with a world-class industrial infrastructure: railroads, banking institutions, internal communications networks, education, mining, shipbuilding, and weapons manufacture (Crawcour 1988; Huddle and Reich 1987; Iriye 1989; see Sansom 1964). Ownership of these facilities originally fell to the central Meiji government. But in the 1880s, when the state discovered to its chagrin that most of its operations were running at a loss, it sold less strategic enterprises to private citizens and relatives of those holding the government's reins. As Huddle and Reich (1987) point out, the conditions of this privatization led to strong ties of reciprocal obligation between state and industry. Japanese entrepreneurs found not an inhibiting regulatory state apparatus but a government eager to promote rapid industrialization with favorable laws and preferential contracts, loans, and subsidies (see Crawcour 1988). This early collaboration between industry and the state set the stage for a century of cooperation toward the goal of growth and national advancement.

Social relations in Japan had long been shaped by Confucian ideology that mandated a patriarchal model of society. From the Meiji Restoration onward, this preoccupation with family ideology intensified and betrayed certain concerns over purity broached in Chapter 6. The promulgation of the Meiji Constitution (1889) explicitly positioned Japanese as subjects of the emperor. The Imperial Rescript on Education (1890) was intended to instill in schoolchildren filial piety and loyalty to the state. And the Meiji Civil Code (1898) made the patriarchal family into the "basic unit of the ruling order of the state" (Miyake 1991, 270). As Louise Young writes in her authoritative *Japan's Total Empire:*

> Enshrined in the Meiji Constitution and disseminated through primary-school textbooks, the concept of a family-state rested on the mythohistorical account of a common racial ancestry. Classical texts traced the origins of the Japanese people back to the divinely descended Emperor Jimmu...mythic progenitor of what was called the Yamato race. In these terms, the nation constituted an extended family linked by blood and the imperial head of state was also the paternal head of this extended national household. (Young 1998, 366; see Miyake 1991, 270)

This conceit of purity of blood pervades the discourse of the period and mirrors attitudes toward microfamily purity (discussed extensively in Chapter 6) as maintained, for example, through nuptial politics and female sequestration. Though this patriarchal metaphor surfaced in articulations of political power, its rhetoric was also appropriated by Japanese industrialists and architects of modernization, who discerned that the social pressures of industrialization required stronger ties between workers and managers. Many workers who had migrated from rural areas to centers of enterprise were maltreated in their new positions of employment, and Meiji leaders became concerned that these conditions might imperil their plans for national advancement (Hane 1998).[2] As Huddle and Reich (1987, 44–45) explain:

> Business leaders sought ways to ensure their workers' loyalty and obedience. They soon found the answer in the traditional structure of the family....Management undertook a deliberate campaign of praising the feudal family structure and the virtues of using it as a model for employer-employee relations. The company came to be regarded as one vast family with management playing the benevolent father role and the workers accepting the submissive role of children....Individual freedom and independent action were to be willingly surrendered for the common good.

A further imbrication of family and industrial growth developed during the second half of the nineteenth century, for amid this top-down repositioning of the family in the structure of Japanese society came a sustained drive to increase procreation. The Meiji vision of Japan as a powerful industrial nation required large numbers of workers to fill factories and men to fill the ranks of the military, so abortion and infanticide were outlawed (Bowring and Kornicki 1993). Probably due to improved living conditions, the nation's population swelled. Along with the mid-Meiji campaign for industrialization—encapsulated in the slogan "rich country, strong army" *(fukoku kyōhei)*—came Japan's promotion of women's reproductive role in preserving the family-state system, which was reflected in another slogan, "good wife, wise mother" *(ryōsai kenbo)*. An analysis of the state's management of women's productive and reproductive roles, historian Yoshiko Miyake (1991) argues, reveals a strong preoccupation with national fecundity. This pronatalist orientation only deepened after the 1920s, when industrial development was seen to have weakened the *ie* household system both as a kinship organization and as a labor organization for agricultural production (Miyake 1991). Average ages of marriage for men had risen during this time due to conscription, expanded higher education, and urban migration (Garon 1997), a perplexing statistic for the progrowth, population-focused leaders of the time.

In response, where Meiji architects had promoted the legal power and moral authority of the household patriarch, Shōwa ideology embraced imagery of maternal fecundity and warmth to buttress the family system. Government propaganda exhorted women to "'Be Fruitful and Multiply for the Prosperity of the Nation,'" and a popular image portrayed soldiers prepared to fight well and die calmly, encouraged by their mothers (Miyake 1991, 271; Hauser 1991). Women's groups urged girls to "'move from an individualistic view of marriage to a national one and to make young women recognize motherhood as the national destiny'" (Havens 1978, 135). Awards were given to fertile mothers who had produced large families (Miyake 1991). A 1941 "Outline for Establishing Population Control Policy" pushed numerous policy ideas that flourished in this pronatalist climate (which would no doubt bring a tear to the eye of any contemporary Japanese bureaucrat tasked with boosting Japan's anemic fertility rate). Public bodies were to encourage matchmaking; girls' schools were to teach childbearing and hygienics; employment conditions were to spare potential mothers and reduce any hindrance to early marriage; single persons were to be taxed higher than married persons; large families were to receive preferential treatment; pregnant women and their babies were to be protected by an improved maternity hospital system; and contraception and abortion were to be further discouraged, as were venereal diseases (Miyake 1991, 278–279).

Discourse linking purity, virility, and national strength threaded throughout the "family-state" propaganda and state policies of the time. Japan's colonization of the Asian continent is a case in point, as Young's excellent study again illustrates. The Great Depression of 1929 had created something of a crisis in rural Japan because after decades of population increase, large numbers of newly unemployed urban laborers had returned to villages to work scarce farmland. This was ready fodder for expansionists. Advocates of Manchurian colonization argued that overpopulation had "'made Japan putrid, like stagnant water,'" whereas the limitless, fertile plains of Manchuria could provide a means to "'purify'" Japan (Young 1998, 327, 325–326). This fixation on Manchuria echoed views expressed by leaders as far back as the early 1900s, who had urged Japanese settlement that could double or triple the area's population (Matsusaka 2001).[3] Recruiting literature invoked themes of blood and family, stressing that "'the rich soil of the continent beckoned' especially to second and third sons, whose 'bleak expressions registered their despair at making a life for themselves on the family farm.' In similar fashion, the language of the village colonization program literally sig-nified the creation of a 'branch village' *(bunson)*" on the Manchurian frontier that would be connected to the Japanese village at home (Young 1998, 367, 336–338).

Emigrationist propaganda represented the colonization of Man-churia "in the figure of a virile young man" (Young 1998, 367). Anthems praised "the physical vigor and power...the 'capable fists' and 'strength' of youth...'stalwart sons of the holy land'" (Young 1998, 367). This was an agrarianism that served the empire; Japanese ministers exhorted "'soldiers of the hoe' to 'Go! Go and colonize the continent! For the development of the Yamato race [*Yamato* (sic) *minzoku*], to build the new order in Asia!'" (Young 1998, 364). This invocation of virility and youth, this national exuberance, had been building for some time. Japan had long fancied a role for itself in East Asia, portraying itself since the 1870s and 1880s as a "youthful and progressive Japan guiding aged and decrepit Asian neighbors down the path to Western-style civilization and enlightenment" (Young 1998, 367). This image would, in time, segue into that depicted in a World War II Japanese cartoon, in which fat, aged, and weak Winston Churchill and Franklin Roosevelt step into a box-ing ring to save a battered Chiang Kai-shek, only to be given a drubbing by a young, strong, pure, virile "Japan" boxer (Dower 1986, 198–199). Appropriation of Chinese territory was justified in terms of the family-state ideology: Manchuria was depicted as a "'brother country to which Japan had given birth' *(Nihon ga unda kyōdaikoku)*" (Young 1998, 366). Indeed, the colony was characterized as blood kin in the persuasive rhet-oric of hierarchical kin relations that originated in village Japan: "'Japan is the stem family and Manchuria is the branch family'" (ibid.).

Hand in hand with the strength and virility of the Japanese male settler was the fertility of Japan's women: indeed, "drawings and pho-tographs of settler women seemed always to catch them with a baby at the breast" (ibid., 369). Involvement of the state in Japanese procreation continued on the frontier as it did in the homeland. While pronatalist colonial authorities had originally encouraged only young settlers to emigrate to the new land, high rates of miscarriage, premature birth, and infant mortality in time changed their minds. Settlers were urged to "'take the old folks with them' to help with this second generation" (ibid., 369, 370).

The course of the war did little at first to discourage state intervention in the sexual life of its citizens, but it put this edifice of the family-state on precarious foundations. Despite the insistence of the prime minister's wife, Tōjō Katsuko, that "'having babies is fun'" (she was a mother of seven) and urgings that women be frugal in order to channel their resources into raising large families, the government's pronatalist campaign was largely unsuccessful (Havens 1978, 136–137). Wartime births decreased due to the tougher times (including public health problems) that came with Japan's slow military decline starting around 1943. However, Havens (1978) notes that Japan's population in 1950, even taking into account the three million war dead, was largely unaffected by a war whose bloodiest period lasted only about half a decade—a relatively short period by demographic standards. With Japan's prolonged retreat in the Pacific, the wartime logic of production and reproduction underwent increasing strain.

These pronatalist wartime policies developed out of long-standing social attitudes toward procreation that betrayed intense preoccupation with purity. Indeed, despite pervasive rhetoric invoking Japan's "'one hundred million hearts beating as one'" (Dower 1999, 122), the population remained governed by hierarchy and segregated with an eye toward pollutant human elements. The state had licensed prostitution for decades in an effort to protect society—and particularly the "'daughters of good families'"—from victimization and contamination (Garon 1997, 100, 103). Prostitution was seen as a way to preserve the family system by providing an outlet for male sexual urges; at the same time, only females were punished for adultery, in order to quash female sex roles deemed subversive to the family ideology of the time (ibid., 101ff.). Though some groups during the prewar period advocated abolition of licensed prostitution, the terms in which they did so urged "purification of blood" *(junketsu)* and elimination of venereal disease for healthy recruits and healthy mothers, notions characteristic of the era's nationalistic, eugenic ideas (ibid., 110). Throughout the war, the state had been reluctant to subject nubile, fertile young women to the rigors of factory work, but by 1944 wartime authorities were forced to rewrite their earlier pronatalist blueprints to make up for labor shortfalls. Nonetheless, the state remained uncomfortable with what Miyake (1991, 281) calls the "incompatibility of women's reproductive roles and their working life" within the ideology of the Japanese family system.

Japan's defeat and surrender at the end of the war came as a severe blow to the country's cultural-nationalist ideology. The emperor's forced capitulation, broadcast widely throughout Japan on 14 August 1945, acted as a symbolic decapitation of the Japanese family hierarchy at whose pinnacle he had sat.[4]

Defeated Japan made early attempts to protect the purity of its procreative base. A matter of days after the surrender, prefectural governors mustered together tens of thousands of "fallen women" (mostly prostitutes) to service the sexual needs of the occupying Allied forces to prevent widespread rape and were told to discourage male Japanese from consorting with the women to maintain a racial "hygienic buffer" against venereal diseases (and racial miscegenation) (Garon 1997, 111, 197; Bix 2000). A member of the Ministry of Finance, citing the amount budgeted for the project, was quoted as saying that "'a hundred million yen is cheap for protecting chastity'" (Dower 1999, 126). In a probably not coincidental convergence of figures, the associations charged with running this tributary of the "water trade" invoked "the great spirit of maintaining the national polity by protecting the pure blood of the hundred million" (ibid.), citing the rough, rhetorical estimate of the Japanese population invoked by war ideologues.[5]

The postwar period became an unanticipated Darwinian interlude for war victims, displaced families, despised war veteran returnees, neglected orphans, all-but-abandoned war widows, and atomic bombing survivors, an underclass that constituted "the country's new outcasts" (Dower 1999, 61). (Recall that Horiuchi's own de facto "untouchable" district, Oden Peddlers' Alley, was settled after the war by Tokyo homeless.) But as every society's choices betray often deeply held social attitudes, Japan largely opted to ignore such suffering in favor of other priorities.

Postwar prostitution perplexed Japanese elites in particular, and the issue of Japanese women consorting with American soldiers during and after the Occupation was characterized by a leading doctor of the time as a "'eugenic problem of mixed blood.'" He warned that "unions with foreigners resulted in a 'difficult delivery' for the mother...so 'chastity must be prized for the sake of our descendants'" (Garon 1997, 200). Unfortunately for control-oriented Japanese leaders, however, the Occupation stripped the Japanese state of many of its previously held powers of "moral suasion," to use the expression that Sheldon Garon (1997, 150–151) memorably employs. The Supreme Commander of Allied Powers (SCAP) limited the administrative functions of neighborhood associations, dismantled wartime "patriotic" associations, outlawed state support and supervision of Shinto shrines, prevented private groups from performing governmental functions, and disbanded the Home Ministry itself, among other measures. As a result, SCAP managed to rend apart much of the "technology of nationalism" that Japan had used so effectively since the early 1900s (Pyle 1973, 53). Neighborhood associations did, however, remain a force after the Allied Forces left Japan in 1952.

But the dominant theme of the postwar period was economic growth, and local groups focused on boosting household savings and encouraging household austerity measures, both intended to husband resources for production-led growth (Garon 1997). Despite attempts to cleanse immoral sexual influences from Japanese social life—women's groups campaigned zealously against prostitution, particularly until the 1950s, and tried to instill a "'new morality'" in Japanese youth and "'foster concepts of purity,' 'encourage healthy entertainments,' and 'eliminate harmful publications, movies, and other materials'" (Garon 1997, 201)—the postwar period was characterized by far less intervention in the sexual activities and procreative plans of Japanese families. Due to this lack of influence and the increased expense of raising children in Japan, among other factors, national fertility began a fall that the state has been conspicuously unable to cushion or reverse.

Births of a Nation: Toxic Waste, Infertility, and Decline in Contemporary Japan

While Japan's long-standing thirst for growth sprung anew in the decades following the Allied Occupation (see O'Bryan 2009), the nation's birth rate, largely decoupled from this project, began a long decline that continues in the present day. Whereas (frequently urban) postwar invocations of the social warmth and fecundity of Japanese village life have long seemed a discursive stretch, now the feeble national fertility rate has become something of a collective despair, a demographic sword of Damocles hanging over the Japanese nation.[6] Given the plunging birth rate and aging population, a pension disaster looms as fewer young workers struggle to support a growing pool of retirees. This foregone eventuality weighs heavily on contemporary Japan: the term that describes the trend, *shōshi-kōreika*, was commonly on the front pages of Japanese mainstream newspapers and on the covers of magazines throughout my fieldwork. One popular daily, the *Asahi Shimbun*, dedicated a weekly front-page column to the subject starting in 2000.

This natalist pessimism intermingled with a pervasive gloom regarding Japan's economic prospects during the time I conducted fieldwork. Indeed, so widespread was Japanese dejection over the economy that it became all-but-omnipresent in my interviews and discussions regarding a wide range of social issues. The unfaltering economic growth that had buoyed the nation's accounts and its spirits during the prosperous postwar decades had ground to a halt (e.g., Bank of Japan 1998, 1999). In its place, a twenty-year-long, on-again-off-again recessional economy choked growth down to a trickle, with colossal bank debt and other fiscal hindrances impeding attempts to resuscitate the nation's financial system.

It is difficult to overstate the effect this downturn had on a Japanese society structurally oriented toward and accustomed to growth. For example, Kumi-san, describing the apparently fluctuating prices some accused Mama-san of charging at Chikurin, compared them to the troubled Nikkei stock index: "Sometimes it's up, sometimes it's down. Who knows why....It depends on the day." The wry fatalism implicit in her tone reflected the bad times that had seeped into everyday life, even to the local bar. One indication of the toll on Japanese society lies in the suicide figures: from 1997 to 2000, more than 30,000 Japanese killed themselves per year, or nearly 100 people a day. As Gavan McCormack (2002) observes, this is three times the number of traffic fatalities in Japan. Along with those who could not find stable work at all, a greater proportion of the workforce shifted to part-time *(pāto)* work—as we see later in the case of Tōda-san—not infrequently juggling several jobs at once to make up the income shortfall. Significantly, part-time work usually involved a concomitant social demotion. "Part-time" is as much a hierarchical distinction as a description of hours worked, since many part-time employees work full-time but are not entitled to full-time benefits, including job stability. While the number of Japanese with "lifetime employment" *(shūshinkoyōseido)* has long been considerably lower than commonly believed by foreign commentators, it took a sharp dive as companies struggled to stay afloat. In the late 1990s, for example, major companies slashed their intake of college graduates, depriving many of them of the raft of privileges that were once an expected part of the salary package for qualified candidates. At the same time, many employees were encouraged to take early retirement, sometimes through heavy-handed coercion and bullying.

Moreover, to the chagrin of adults invested in reproducing Japanese values such as hard work and sacrifice in coming generations, Japanese who reached working age during this period sometimes had other plans. There have long been young Japanese less enamored of establishment, mainstream Japan, and economic causes and scarcity of "good jobs" cannot be ignored. Yet during this downturn, some graduation-aged Japanese began to take flexible, low-status work with, effectively, no clear future prospects in order to pursue a dream or just to live a freer life: to become musicians, artists, fashion designers, hair stylists, and writers, or, simply to work somewhere hip rather than in a machine shop or factory.[7] Known as *frītā*, a neologism combining the English word "free" with the German term for worker, *Arbeiter*, these relative hedonists constituted an increasingly mainstream subculture, full of affective disenchantment with the system and ill-defined aspirations (Genda 2004; C. Smith 2006). Once off the traditional employment path, the social stigma of being a *frītā* made it harder to find other work or—for males—to find brides (C. Smith 2006). In an overlapping category, legions of employed young women and young men opted to live at home with their parents far longer than was previously considered appropriate, sometimes well into their thirties—this in a society where traditionally women "had" to be married before the age of twenty-five and men often became independent in their twenties. Known as "parasite" singles *(parasaito)*, these young people postponed, or at times rejected outright, the prospect of marriage and childrearing in favor of a longer liminal social status and the freedoms that came with it. In the end, of course, plenty of Japanese "parasites" chose eventually to start families, but their choice to do so later, coupled with economic concerns, made improvement in the birth rate a near impossibility (see Retherford, Ogawa, and Matsukura 2001). Informants consequently had little confidence that Japanese society would turn itself around in time.

The social ramifications of economic stagnation were palpable and widespread both in my fieldsites and in Tokyo as a whole. Since years of declining growth had developed into recession, local businesses either retrenched aggressively or, in some cases, went under. In Horiuchi, the effects on the shops along the main shopping street were devastating. While many of these were family-owned by long-time residents, and were therefore spared crushing rental costs, nearly all vitality had disappeared from this local artery. For example, Tanaka-san, the owner of a sushi shop located on one of the major thoroughfares, complained of a severe drop in customers. In contrast to the robust consumption and good profits of the Bubble period, he had long stretches of only two or three customers a night on average, with somewhat more lunchtime business. In spite of this, he opened shop every scheduled night, bought (and consumed) raw fish and other culinary supplies, and spent most of the evening watching television. "I could probably go catch the fish

myself," he joked ruefully one night. Tired-looking local electronics outlets, an inordinately high number of which seemed to be located near
each other in a quiet area, almost always stood empty. A traditional
sweets shop rarely had customers, and the three rice stalls spread along
the length of the gently winding street had only a comparative trickle of
customer traffic over which to compete, with regular clients defecting to
buy in supermarkets or other bargain outlets at a discount. Such small
local merchants, who depended on customer loyalty based on chit-chat,
trust, and relationships cultivated over time, found that visits became
less frequent or that customers bought less each time. One local sun-
dries grocer lamented, "Maybe I should quit and buy stock in Super-
Mart," a large and successful (pseudonymous) bargain outlet located
an easy bike ride away. Exceptions included Mama-san's bar and restaurant Chikurin, a rare beacon of social mirth and prosperity amid the
gloom, and Ishimoto-san's thriving recycling business, which I examine
in Chapter 8.

Concerns over the economy dominated media coverage and local
discourse. While "growth" had stood as the central metaphor for the
boom times, articulations of adversity were more varied. In the media,
climatic metaphors reverberated in such terms as *teimei* ([clouds] hanging low) and *hiekomi* (a cooling). *Chintai* (stagnation, dullness, inactiv-
ity) and *teitai* (stagnation) communicated the undesirable inertia that
was the obverse of robust growth. *Kotai* (retreat) lent a military impression to receding economic vitality, and *ochikomi* (a falling in, a caving in)
resonated with the most prevalent metaphor for discussing the swift end
of the Bubble years: *hōkai* (collapse). Mama-san, for her part, preferred
the dramatic *hajikeru* (to burst apart). In local discourse, terms such as
fukakujitsu (uncertain, unreliable), *fuan* (uneasy), and *fuantei* (unstable)
were commonly aired during this time, reflecting the negativism of the
times not only semantically but syntactically, with the negative prefix *fu-*.
This they shared with by far the most common term for the change in
fortunes in Japan, *fukeiki* (bad times). In conversations with informants
in Horiuchi, *fukeiki* seemed to go beyond the economic to describe the
whole post-Bubble hangover that afflicted the Japanese I interviewed.
So closely was the economic scene imbricated over the social in Tokyo
that *fukeiki* seemed to index not only the downturn in fortunes of many
locals and the value of their holdings (particularly small landowners) but also a change perceived by some in the values of members of
the community. Mama-san's frequent comments regarding the lack of
heart *(kokoro)* of people on the main shopping street was consonant

with this theme. In her opinion, if everyone helped each other, according to the community ideal so often invoked explicitly and implicitly in contemporary Japan (Robertson 1991; Bestor 1989), business would be better. She explained: "Now it's bad. But it used to be good. If people had heart *(kokoro)*, if people communicated, they would have a lot of business....Cars don't stop now, but if it was genuine *(shinsei)*, the cars would stop. Everyone's thoughts are scattered....[Everyone's] always bad-mouthing [each other] *(itsumo, warukuchi bakkari)*." The cancer of local *shimaguni konjō* back-stabbing,[8] evident in fissures between community groups and even within groups, tended to metastasize during the bad times, as I illustrate further in the next section.

The steady encroachment of urban ills gave little comfort that Japanese society was heading in the right direction or that life would improve. Indeed, reports on toxic waste and similar chemical perils pervaded Japanese mass media during this time, making the topic nearly impossible to avoid. Since ideas of nature are, in many ways, central to Japanese self-imaginings, contaminated landscapes, polluted cities, defiled seas, tainted fish, and climate change served to deflate Japanese optimism further. The malaise that gripped my Japanese informants frequently seeped out of this complex of pessimistic news and dreary social surroundings, of which toxic waste threats, flagging fertility, and anemic business prospects were important elements. Frequently, these concerns came refracted through the prism of reproductive challenges.

Toxicity, Reproduction, and Identity

Mrs. Fukushima, the mother of two teenagers, spoke with me in a quiet Horiuchi café at the tail-end of the Tokorozawa dioxin scare. "That incinerator nearby, right? The terrible thing is, people living around there talk about babies being born with only one arm or fingers missing. There are miscarriages. Cancer....I don't know what I'd do [if I were having another child]." Of course, birth defects, miscarriages, cancer—these misfortunes befall families all over the world in every conceivable living environment. But the fact that they were being interpreted by some Japanese in terms of toxic incinerator smoke in residential communities was highly significant. Mrs. Fukushima's claims regarding dioxin emissions from the large incinerator in Sugibayashi were not unusual. Mass-media coverage of dioxin-contaminated breast milk throughout urban Japan, and other lurid stories, had by this time intensified the reproductive register of environmental anxieties in Horiuchi. Of course, Mrs. Fukushima and others who made such complaints did not generally know more than a handful of people actually living in Sugibayashi; though such anxiety was mirrored in Sugibayashi among residents there, living as they were in the shadow of an enormous smokestack and the facility's disturbing operations, in Horiuchi these claims tended to be both vaguer and more inflated and took the form of rumor that could not be (and was not expected to be) confirmed.[9] Significantly, I heard of no strikingly unusual birth defects, nor any dramatic increases, occurring in Horiuchi during my fieldwork.[10] But dioxins in particular were a malleable trope, and articulations of dissatisfaction that involved health problems often included dioxins and air pollution more generally as vehicles of blame for such misfortunes.

We have only to take the unfortunate example of miscarriage. Nine of my informants admitted they had suffered miscarriages, and others knew of further incidents both in the area and elsewhere, though the latter group declined to discuss these cases in much detail.[11] One older informant, Enguchi-san, had married when she was a young woman into a family that ran a moderately sized service-sector business. Enguchi-san's comments make it clear how sensitive reproduction can be in Japanese families, though hers were perhaps atypical circumstances. Her new husband had four sisters, and life was not made easy for the (paradoxically) charismatic and engaging Enguchi-san: "I worked so hard, and yet my sisters-in-law criticized me constantly. I became pregnant before long and continued to work tough hours while carrying the baby. Twelve-, fourteen-hour days. Still, they spoke ill of me, no matter what I did.... When the baby miscarried they all blamed me for it.... Eventually, I decided to divorce my husband...for a variety of reasons." While each family differed, childbirth in Japan was a sensitive undertaking and could become a flashpoint for women viewed as being insufficiently careful, fertile, selfless, or other cruel such judgments.

Tōda-san had been married to his wife for over twenty years when I met him. They had tried to have children, he explained, but after his wife's second miscarriage, they gave up. Tōda-san's voice lowered when he spoke of this sorrowful chord running through his marriage, and it was obvious that discussing the subject was difficult for him. Asked what he thought might have contributed to their reproductive complications, he paused, sighed, and murmured, "I don't want to know the cause." Then, after a moment, he continued:

> I can't say that [dioxins] definitely did have an effect, but I certainly wouldn't say they didn't have.... It's like that [artificial sweetener]

that they said causes cancer some years ago. There's no way to prove that there's a direct effect. But most likely, these chemicals have their bad effects. They're bad for humans, bad for animals....Dioxins are bad. They've been around for awhile, over thirty years. First in the Vietnam War with the American Army, and then now with the harvests in Japan. They can cause cancer. Genetic problems. Transmitted by sperm.[12]

Unfortunately for Tōda-san, other male members of the community openly derided his masculinity. This resulted partly from Tōda-san's personality and life choices: Tōda-san was wont to be somewhat nonconformist in conversation, appearance, and demeanor, by middle-aged Japanese standards—he wore a goatee in his late forties, and behaved in a vaguely bohemian manner—and he performed odd jobs to pay the bills. He had also dedicated himself to photography as a part-time vocation. But a running theme in their mockery, particularly that of Katsu-san[13] and some of the other middle-aged men, was that he was boyish, immature. One night, for example, they made a point of teasing him at great length when he mentioned that he had gone to a Self-Defense Forces parade to take photographs. They laughed, saying that he wanted to be a soldier, just like a boy would.[14]

The subtext was that Tōda-san was a married male in his late forties with no children. Indeed, many of his later choices of career and leisure activities derived indirectly from the fact that he did not need to support a family. He and his wife both worked, but unlike some of those who mocked him, he had a relatively fluid schedule and could pursue occasional jobs that he enjoyed, such as freelance photographic work. It was clear, however, that this freedom came with a social price. Tōda-san and his wife would have dearly loved to have children; yet their absence, and the absence of parental obligations, undermined his ability to be taken seriously as a "real man" by his peers.

Sizing Matters: Anxieties over Sex, Virility, and Manhood

Another, more direct manner in which dioxins were intertwined with perceived manhood was first voiced to me privately by women in the community. Japanese learned via media reports and other sources—including gossip—that dioxins impede the development of the male genitals (see Roman and Peterson 1998). Though few men discussed this effect of dioxins without prompting, their silence was telling. One Health and Welfare Ministry official, who was reading aloud from a ministry report on dioxins during an interview—and assumed I could not read scientific jargon in Japanese—actually skipped over a sentence mentioning this side effect of the toxin. When I pointed out the omission, he took up that symptom too, but in somewhat guarded tones. I should say, however, that our conversation took place in an office full of Japanese men with some female staff members within earshot. The discomfort of Japanese males with the topic is hardly surprising, and an exaggerated belief that "bigger is better" is certainly not peculiar to Japanese men. (Revealingly, though technoscientific studies have determined that only testicle size is affected [e.g., Ohsako, Miyabara, et al. 2001], much of the discourse surrounding the issue mistakenly assumed that penis size was the casualty of dioxin exposure.)

Two *débauchés* well beyond the age of being embarrassed by such matters were among the few men who raised the topic spontaneously. My friends Shin-chan, a construction worker, and Taka-chan, a sushi master, sat for an interview together one night. Asked about toxic pollutants in Japan, they replied,

[Taka-chan:] The most worrying thing is environmental hormones *(kankyō horumon)*. How do you describe environmental hormones...Sex—well, look, sperm count completely goes down, that's how it seems, anyway *(sekkusu ga, hora, seishi ga hecchau toka, sō mitai yo)*. So people's sex weakens...

[Shin-chan:] Like people who lose their power, virility *(pawā ga naku naru hito toka)*....

[Taka-chan:] Take shellfish *(kai)*. The males, well, their reproductive organs shrink way down. PCBs, polychlorinated biphenyls[15]...you know, they spill into the sea from the toilets of ships, and the male reproductive organs of shellfish gradually shrink down *(kai no ne, osu ha, anō, seishoku-ki ga dondon hete kichau to iu no ha, PCB...are ha soko fune no soko toire wo tsukatte, are ha umi ni nagarete, kai no osu no seishoku-ki ga dandan hete kiteru)*....

[PWK:] So are the effects the same for humans as for shellfish?

[Both:] Yeah, yeah [definitely]....

[Shin-chan:] I mean [Shin-chan gestures around at the building and neighborhood], look at all the chemicals around here, like glues *(sec-*

chaku-zai) and building chemicals *(kenchiku-zai)*.[16] ...We've got the worst air pollution, right here in [Horiuchi].

Though these two men seized immediately on the sexual repercussions of toxic waste, others might only mention them over time, and reluctantly. Furthermore, it is significant that both men, neither of whom was white collar, were exposed through their professions to surprisingly detailed and accurate specialist knowledge about toxic waste.

Long-term exposure to dioxins and other endocrine disruptors can effect a feminization of male anatomy and sexual behavior, as well as decreases in testes weight, sex drive, sperm count, and ejaculation (see EPA 2000, esp. chap. 5),[17] so Shin-chan and Taka-chan's description of toxic effects was relatively close to the mark. Japanese male anatomy was an extremely sensitive topic in Japan, and it was not easy for a Caucasian male anthropologist to tread such controversial territory. Japanese are well known among visiting social scientists and others to engage in discourse about what they see as foreigners' relatively larger endowment[18] and apparently undisguised sexual interest in Japanese women. Furthermore, because not a few Japanese women seem, particularly to Japanese, to be attracted to some foreign men, this topic tended to bring out very strong opinions from some Japanese males.[19] Some interviews were particularly illustrative in this respect. When discussing the (barely, by U.S. standards) deteriorating security situation in Tokyo, Andō-san explained, "Asian *(ajia-kei)* foreigners are dangerous because they commit crimes [and are hard to catch because they look Japanese], but Westerners—Westerners are bad because [they go after] women *(seiyōjin ha, seiyōjin ne, onna ni warui)*." At the same time, he made the Japanese hand sign for a nubile girl, an extended pinkie finger with fist facing palm down. Despite Andō-san's grumpiness, one reason I found fieldwork relatively easy to accomplish in Horiuchi was, ironically, that my "foreign" long-distance relationship status was well known there. One Japanese woman, Miwa, who had a Japan-based ten-month relationship with a Caucasian foreigner, brought up at least one reason why interracial romance might have been seen as a problem: "When I went out with [the foreigner], every single boy I knew commented on [his size]. They would say it in different ways, but *every single one* did. If I had to tell them, I would say, 'It doesn't matter.' But I don't think they would even believe me if I did." At one extreme, then, we have Andō-san, consumed as he conspicuously was with monitoring racial sexual boundaries, with distinct echoes of Japanese purity-obsession. At the other, we have Nakamura-san, a well-traveled, urbane, middle-aged Horiuchi resident, avowedly fond of white *(hakujin)* female foreigners. When asked why he did not marry one, he replied (in English): "I can't give a Western woman what a *gaijin* can give her." The scientific particulars of the effects of dioxins and endocrine disruptors were murky and ill-understood by most men, but these attitudes, taken as a whole, hint at the sensitivity surrounding these issues and why the genital and sexual threat of toxic waste was taken seriously.

The responses of female informants were frequently more candid and revealing. One woman in her late fifties relayed what she had heard about the effects of dioxins and other toxic waste from the obstetric nurse mentioned above: "Maki has some scary stories....She delivers babies everyday, right?...Maki was saying that the genitals *(seishoku-ki)*[20] of baby boys are getting smaller! She's been working there for six years, and she says [there's been] a difference from when she started till now!" Another woman, a friend in her sixties, reiterated this, umprompted, in a general discussion of dioxins: "[Maki-chan says] babies' genitals are half the size they used to be. Really, they're small. *(Seishoku-ki ga, mukashi yori, hanbun. Chīsai no.)* That's definitely related to dioxins....So genitals, male babies' genitals, are only half what they were in the past....Of course [it's related], of course. It came out in the newspaper, right?" When I brought up the subject in interview with the nurse, Maki, a week later, however, she denied any knowledge of such a thing, and her defiant facial expression and manner made it very clear that she simply did not want to discuss the topic. Another jocular middle-aged woman whom I had gotten to know well commented, without prompting, "You know, Peter-san, this isn't talked about much, but [they say] that dioxins make [male genitals] smaller . . ." She then went on to talk about government policy regarding dioxins, during which she castigated the laxity with which the issue of toxic threats to health had been handled over the decades. While I am sure that some of the other people with whom I gingerly brought up this topic expressed genuine ignorance of these rumored genital ramifications of dioxin exposure, others—like the aforementioned Health and Welfare Ministry bureaucrat—likely preferred not to discuss the subject.

The social impact of these data was enhanced by other information circulating at the time. Media reports speculated whether the sperm count of urban Japanese males was in decline (a notoriously imprecise and politically charged calculation; see Swan, Elkin, and Fenster 1997); other reports dwelt on male sterility *(dansei funin)* and possible treatments through traditional Japanese medicine *(kampō)* (see Baba, Iwamoto, et al. 2000; Tsujimura, Iwamoto, et al. 2000). Sitting on a park bench by the river, watching junior high school baseball practice nearby, my older female friend from the previous paragraph made reference to what had been in the news, showing how such information could be quickly exaggerated: "Already sperm count is low *(Mō seishi no kazu ga sukunai)*. And if [dioxins] are doing that, and if we have [male] genital impediments, there'll be fewer children. And people now, it's not that they're not giving birth but that they *can't* give birth *(De, ima no hito ha umanain ja nai no yo, dekinai no yo.)*. It's terrible. So there are fewer schools, fewer students. [Pointing at the nearby school] [It's] like that in Horiuchi, too."

The state responded—in something of a fit of pique by Japanese bureaucratic standards—in December 1999 with the results of a wide-ranging Japanese Environment Agency study. It determined that Japanese sperm quality and testicle weight had not decreased significantly since 1964 (Japanese Environment Agency 1999b).[21] Against the backdrop of a well-publicized declining birth rate, this gossip and reporting lent a new and worrisome anatomical-biomedical dimension to discussions of Japanese procreation that in the past had tended to dwell on socioeconomic causes, such as later marriage, the expense of raising children, and so on. Along with the contemporaneous anxiety over dioxin-contaminated breast milk, it was as if everything natural was being disrupted, corroded, undermined.

Another dioxin-related problem disturbing to some of my informants involved sex birth ratios. According to extensive technoscientific research, exposure to dioxins can lead in time to fewer male births. Japanese have historically prized sons, and though this particular effect of dioxin was less well known, once news began leaking out, it was viewed gravely, to say the least.[22] Beyond the level of illness or environmental damage, then, toxic pollution had the potential to magnify anxieties surrounding reproduction and identity in a society well aware of its low level of procreation compared to idealized days of yore. What had seemed like a localized problem—environmental pollution—had become a blight threatening current generations of Japanese and their future descendents and thus the very nation itself. In Horiuchi, nevertheless, these anxieties did not stimulate organized protest. One woman railed against the lack of protest regarding what were, to her, critical issues: "They [her neighbors] should be worrying about them with louder voices. All they say is, 'That's too bad,' when they hear about [environmental problems], but those people's children's children, it'll be bad for them, too."

Significantly, these reproductive ramifications of toxic waste came to light while Crown Prince Naruhito and his wife Masako, married in 1993, were taking a long time having their first child, a highly publicized issue inside and outside Japan. Most Japanese coverage was deferential and hopeful, while foreign media began making insulting accusations (e.g., *Süddeutsche Zeitung Magazin* 23 February 2001, for which the publication later apologized; see *Independent* 17 April 2001). In time, the imperial family began consulting fertility specialists, and the princess became pregnant in 1999, but then miscarried—devastating news for the couple as well as the nation. Her next pregnancy, also believed to have been assisted, was carried to term and in 2001 the princess gave birth to a girl, which delighted the Japanese but still left imperial succession in question.[23] (Former Prime Minister Koizumi Junichirō, then a candidate, gushed, "I hope this will trigger a baby boom in Japan" [*Independent* 17 April 2001], but such hopes went unfulfilled.) The reproductive travails of the imperial family surfaced in ethnographic interviews and other discourse in my fieldsites. Although, to me, no one in my fieldsites tried to link Japan's toxin-skewed sex birth ratios with the imperial family's situation, the family's problems were raised with regard to Japan's declining birth rate. Given the iconic status of the royals, these difficulties and their news coverage contributed somewhat to the gloom that beset much of Japan beginning in the early 1990s and lingering for nearly two decades.

Of course, while Japanese concern over toxic waste during these years is highly suggestive, dioxins and other environmental pollutants did not dominate the attention of contemporary Japanese in the new millennium. The seriousness and pervasiveness of toxic waste, however, its Hydra-headed nature, and its embroilment in Japan's ongoing reproductive malaise meant that along the uneven, pulsating topography of Tokyo-ites' risk consciousness, toxic waste pulled at the thoughts of Japanese even when not always at the forefront of their minds.

If there was a consistent theme through my interviews in Horiuchi, it was that the Japanese felt angry at or betrayed by politicians and resented or had become cynical enough to expect government neglect of toxic threats. This did not rise to the level of outright protest, but Tōda-san and Enguchi-san made pointed comments regarding the political register of toxic pollution. Tōda-san, clearly upset over the tangle of conditions that have led to what he saw as Japan's pollutant state, burst out:

The government doesn't defend ordinary Japanese against these things. They only think about company relations and economic matters. Now that the Bubble's burst, it's even harder to get the government to think about these kinds of things. In California, they took care of exhaust fumes very quickly. Germany's very strict with these environmental problems. But here it's different. Years ago, [Japanese politicians] knew that dioxins were bad, but since there was no real evidence of it in daily life, they thought it best to let sleeping dogs lie.

Enguchi-san, who had spoken of her miscarriage, said she carried her child while living in a crowded urban community (but not Horiuchi) and that there was certainly pollution in abundance back then. Asked delicately whether she thought such pollution as dioxins could play a role in miscarriages or other birth complications, she replied: "You never know for sure with these things. But it's certainly possible, what with the reports we hear on TV these days.... This is an unhealthy world we're leaving for the next generation—not that the government is doing anything about it." Enguchi-san and Tōda-san were justified in casting aspersions on politicians' past reticence to enact policy change or threaten the close ties between government and business. However, as it turned out, the government was taking the reproductive predicament of urban Japan quite seriously indeed.

Government Attempts at Stimulation

Japan has, in recent years, directly confronted the severity and magnitude of the *shōshi-kōreika* (declining birth rate/aging population) problem. Yet few options remained available to the state to combat the long-standing demographic lag. With the Japanese economy still prone to low growth or even economic decline and with budgets tighter, Japan's leaders chose to focus their energies on shorter-term objectives like economic growth rather than longer-term challenges such as population growth. It was also debatable how much the government could achieve even if it dedicated itself to the task. Nevertheless, an anecdote from Garon's *Molding Japanese Minds* helps express the extent to which Japanese politicians share some of the domineering instincts of their pre-1945 counterparts. Garon describes how in June 1990 then–Finance Minister Hashimoto Ryūtarō and the Japanese cabinet took up the issue of the falling birth rate as well as the rising costs and future labor shortages this trend augured.

> European leaders had voiced similar concerns about birth rates
> in their own countries, but Hashimoto's proposed solution was to
> intervene in the lives of women in a manner that the Europeans
> would not have considered. Reasoning that increasing numbers of
> young women were deferring childbearing to attend colleges and
> universities, the finance minister urged the cabinet to reconsider
> the government's postwar policy of encouraging women to obtain a
> higher education. (Garon 1997, xiii)

The cabinet did not, in the end, take the future prime minister's advice
(he rose to power in 1996) but Garon adds that, concurrent to this decision, the Ministry of Health and Welfare was busy devising programs
to cope with Japan's aging-society quandary. The ministry had spent a
decade promoting the idea, via a willing mass media, that Japan's three-
generation families were best able to care for the elderly, thus sparing
the state much of its anticipated welfare burden (Garon 1997). Though
the Japanese state can no longer employ the "statecraft" (Garon 1997,
xiv) that worked so effectively in the first half of the twentieth century, it can indirectly attempt to achieve some of the same objectives.
To this end, both before and during my core fieldwork in 1998–1999,
the state offered tax incentives, initiated progenitive promotional campaigns, and circulated other reproductive propaganda. These efforts had
mixed results. For instance, as Glenda Roberts (2002) explains, a five-
year solution along these lines, dubbed the "Angel Plan," was initiated in
1994 by the ministries of Health and Welfare, Labor, Construction, and
Education. But this quickly became a burden on communities already
groaning under the weight of obligations toward the growing elderly
population. Moreover, the fertility rate plunged to 1.38 in 1998, two
years earlier than predicted, and an emergency childcare five-year plan
was left underfunded and neglected (Roberts 2002). Still, Tokyo tested
other strategies. The Tokyo Metropolitan Government launched a program whereby young couples could secure free or reduced-price daycare
and education for new children. The prices decreased with each successive child.[24] Furthermore, the Ministry of Labor created a program that
bestowed prizes upon companies with "'family-friendly,' flexible work
policies" (Roberts 2002, 76), an updated variation on the wartime practice of giving awards to parents of large families. Whatever their historical provenance, the policies were welcomed by young couples to whom
I spoke, who were highly receptive to any reduction of the high cost of
raising and educating children in cities in Japan. But these couples also

made it clear that the measures achieved little. No family would plan to have another child based solely on the paltry savings or other incentives made available by the state at this time. One good piece of news for young families came with the historic ouster of the ruling conservatives by the Democratic Party of Japan in 2009. Incoming Prime Minister Hatoyama Yukio promised to pay families a staggering 26,000 yen (about US$285) per child *per month* to lighten the burden of child-rearing (DPJ 2009). However, at the time of this book's completion, it remains unclear how Japan, already overladen with debt, can hope to pay for such an expensive program with its economy still in the doldrums.

It is in this social context that we must assess the state's actions regarding toxic pollution in 1999. One striking development in the wake of the TV-Asahi/Tokorozawa dioxin scare was that national and metropolitan government bodies became aggressive in moving to curtail dioxin pollution.[25] An investigative commission was organized to assess the "tolerable daily intake" of dioxins. Ministries and agencies, in cooperation, set targets to regulate existing incinerators and limit the number of incinerators allowed to operate in close proximity to one another. This move was seen as directly related to unhealthy conditions in Tokorozawa. During the course of the spring and summer, political parties agreed on a tolerable daily intake of dioxins that was 60 percent lower than the one previously urged by the Health and Welfare Ministry. They also vowed a 90 percent reduction of national dioxin emissions, to 1997 levels, by the year 2002.

A Ministry of Health and Welfare bureaucrat I interviewed explained that a somewhat changed political landscape had influenced the shift in policy. The New Kōmeitō political party had become an influential partner in the governing coalition and was troubled by Japan's dioxin crisis. Kōmeitō was formed by members of the Sōka Gakkai, a powerful Buddhist organization in Japan (labeled a cult by critics). The party promotes a "political manifesto dedicated to protecting life and its dignity" (Kōmeitō 2009, 3) and prominently advertises its commitment to reducing dioxin pollution and the effects of dioxins on its official website.[26] Beginning in 1997 in Tokorozawa with small-scale initiatives by local Kōmeitō politicians, the party attempted, slowly, to enact changes in awareness of the dangers of dioxins and, with the TV-Asahi/Tokorozawa dioxin scare in early 1999, such efforts picked up speed.[27] If the extended public outcry in Japan over toxic waste and incinerator safety issues were not enough, Japan came under international pressure after the UN Environment Programme (1999) released its alarming finding that Japan was responsible for approximately 40 percent of all air emissions of dioxins in the developed world.[28]

One relevant policy erupted into something of a minor scandal during my fieldwork and presents an important example of the difficulties the government faces in attempting "social management" in contemporary Japan. In late 1998 the Health and Welfare Ministry provoked uproar when it indicated that Viagra would be approved for use in Japan after only half a year's consideration. Men, in general, welcomed the move, and Viagra became the subject of much lewd discussion and joking.[29] But some women's rights groups and others were furious that Japan, once again, seemed to be pursuing a double standard with regard to male and female privileges—ministry approval for a contraceptive pill, by contrast, had languished for over a decade at the review stage and still had not been granted. Officials for the Health and Welfare Ministry claimed that the safety of the pill was still being determined, though some Japanese women were well aware of the pill's widespread usage in Europe and the United States. Officials also stated that they were concerned about the spread of sexually transmitted diseases, particularly HIV, in a country where these have been kept more or less under control, relative to most developed countries, through widespread use of condoms and other factors—a plausible concern but hardly the whole story.

This male-centric health policy was debated and lampooned in the press and in bars and offices throughout Tokyo. Kumi-san, a generally skeptical Horiuchi-based woman, had this to say: "The Japanese government is usually slow. When they do something fast, you have to be suspicious, don't you." But the policy reflects, upon closer scrutiny, a consistent logic in the state's approach to reproduction in Japan and its concern with the population problem. Any technology that would allow Japanese women to control their fertility would be anathema to a government laboring to increase pregnancies, accidental or otherwise. And though Viagra was perhaps more likely to be used by older men participating in Japan's thriving sex industry (and perhaps even by politicians and bureaucrats involved in the drug's approval), for example, than by men in progenitive relationships with women actually apt to have children, the pronatalist logic was crude but consistent. (The relevant ministry finally approved limited use of the pill in June 1999.)

Clearly, the state had its work cut out for itself with regard to promoting reproduction and population growth. Japanese urban society, during my time there, seemed almost structurally organized to *limit* procreation. The vast numbers of Japanese dozing on public transport, in waiting rooms, and at other visible locations conveyed the extent to which Japanese sacrifice sleep for other pressing obligations; indeed, many wear fatigue as a badge of industriousness (cf. Steger 2003). Sleep deprivation was, for many of my informants, a way of life. The obligation that loomed largest was work. The notoriously long hours many Japanese spend at work—or in whatever organized activity they pursue—left them little time at home with spouses, families, and so on, or with mistresses, for that matter. Couples with both partners working full time wrestled with even more complicated leisure schedules. Furthermore, "play" for salarymen and others is frequently structured. It is not unusual for Japanese full-time employees (usually, but not always, male employees) to spend long hours with work colleagues in entertainment districts. These regular company outings, seen as an essential component of company business, can include negotiations and deals conducted in clubs as nubile hostesses pour drinks, light cigarettes, and fluff up male egos while keeping conversation flowing (Allison 1994).[30] It would be silly to conclude that these factors alone are suppressing Japan's birth rate—after all, while impregnation may often be preceded by intimate drink dates, romantic dinners, and other gallant foreplay, it need not be—but with the widely regarded expense of raising children in Japan already discouraging couples from raising large families, sleep deprivation and absent husbands certainly do not help matters. An informal survey in 2007 by the Japan Family Planning Association determined that, in Japan, 35 percent of married couples had sex less than once a month and that nearly 40 percent of men and women aged sixteen to forty-nine had not had sex in more than a month. The association's director was quoted as saying, "The situation is dismal. My research shows that if you don't have sex for a month, you probably won't for a year" (*Associated Press* 15 March 2007). Added to this, male libido might be aroused (and expended) in "floating world" institutions such as "soap-lands" *(sōpurando)* before salarymen reached their wives at home. To be sure, there are legions of Japanese men who have never been to any such sleazy establishment in their lives, but the thriving entertainment districts catering to largely all-male (frequently work-related) parties in cities all over Japan do indicate at the very least that a significant number of men are being led into culturally acceptable venues for expression of male libido and away from other, more potentially progenitive, encounters.

Furthermore, with easy access to abortion, Japanese women have the ability to control their procreation to an extent that is inconvenient for pronatalist policymakers, to say the least. With the right to abortion enshrined in Japanese law, conservative elements in male-dominated Japanese society, according to Helen Hardacre in her lively *Marketing the Menacing Fetus in Japan* (1997), chose other, subtler means to dissuade females from terminating pregnancies. These included the shrill, manufactured rhetoric surrounding *mizuko kuyō,* Buddhist rites intended to appease the spirit of aborted fetuses (cf. LaFleur 1992). In what Hardacre (1997, 85) calls "heavy-handed, blatant, ideological policing of young women's sexuality that is so obvious in the media blitz on *mizuko kuyō,*" Japanese tabloids and other media preyed upon the sensitivity of women who had had abortions and on the credulity of those who might in the future by depicting the spirit of the aborted fetus as a dreaded, relentless presence haunting the postpartum life of the mother. Hardacre writes, "The victims report a complex of symptoms combining elements of folkloric ghost stories and menopause....They experience 'spotting,'...menstrual irregularity, cramps, headaches, and pains in the shoulders and lower back. Some report sexual problems, such as frigidity or nymphomania, infertility, or strange growths on the genitals," and other hyped symptoms included spousal infidelity, career difficulties, and discipline problems with children (1997, 81–82, 85). These diverse complaints therefore could, and were intended to, apply to nearly anyone:

> For teenagers and unmarried women, the message would seem to be that illicit sex and abortion damage the body and spirit. If you refuse to carry pregnancy (especially first pregnancy) to term, you may not only lose all future opportunity to bear children (the threat of infertility), but also your youth and beauty, the qualities that originally made you desirable as a sexual partner. To married women in middle age, the logic of *mizuko kuyō* says, in effect, "All the problems you now experience with your husband and children are due to abortions you have had. Your husband's infidelity and your children's failures are simply the reflection and consequences of your earlier undisciplined sexuality." To women in menopause and older, the logic says, "There is a reason for the physical ailments you suffer now. Earlier abortions demand an accounting at the end of your reproductive life. You must recompense through ritual." (Hardacre 1997, 82)

Though Hardacre's account focuses almost exclusively on this paranoid, horror-film vision of planned parenthood and its discontents, the editors of tabloids and other lowbrow media are not generally averse to promoting normative views of Japanese women that suit conservative social aims (see Skov and Moeran 1995; see also Moeran 1996). Despite the difficulty of determining cause and effect, and separating clear manipulative editorial intent from "the news" (Bourdieu 1998), it is intriguing to note my informants' impressions that women's magazines, for example, emphasized how the contraceptive pill can stimulate a little weight gain, a clear deal-breaker in a society where young women strive toward an anorexic model of female beauty. This was quite different from the discourse surrounding the pill in Euro-American contexts, which trumpets enhanced breast size alongside greater menstrual regularity, sexual freedom, and reproductive control.

Conclusions

Living in Japan amid a pollution scare and during a recession, it was impossible not to remark the rough symmetry of government efforts to stimulate *economic* growth alongside their efforts to stimulate *population* growth, objectives complicated by toxic waste and by some inevitable frictions between the two. These themes circulated not only at the rarefied level of the national media and state policy-making but also within the social milieux of my fieldsites. Many informants wandered discursively from talk about toxic pollution to talk about Japan's birth rate and vice-versa. And they also moved easily from discussion of (the lack of) economic growth to diatribes against the growing amount of waste and pollution in Japanese society, along with an extensive list of grievances against insalubrious contemporary developments that disturbed them. In the next chapter, I extend this discussion to describe the Japanese state's linked attempts to bring about an environmental paradigm-shift in Japan by (re)constructing Japanese sustainability.

CHAPTER 8

Constructing Sustainable Japan

Black and filthy Tokyo Bay,
Tokyo Bay, where fish cannot live,
What can we do?
We must all take care of you.
Black sea, deceptive sea,
Mercury and sludge-filled sea that died,
You were everyone's treasure...
Oh, how clean your waters used to be.

    —Song composed by a Kawasaki-based fishermen's protest group

In a nation of avid golfers, Wakasu Golf Links has the distinction of being the only golf course in the core twenty-three wards of Tokyo. It is also unusual in that, until not long ago, cigarettes, pipes, and cigars were ruled out of bounds there. Perched on an island in Tokyo Bay, Wakasu forbade smoking because methane and other highly flammable gases diffuse out of the ground beneath the course from the decomposition of the 18.44 million tons of waste interred there (Yokoyama 2007; see figure 14). Indeed, before 1990 Wakasu was known as Reclaimed Lot 15, a destination for Tokyo's waste like its better-known forebear, the notorious Island of Dreams *(yume no shima, also dubbed by cynics the Island of Garbage, or gomi no shima)* (cf. Huddle and Reich 1987).[1] Tokyo, in order to make the amenity of open "island" space available to its citizens, but mindful of the methane danger and the problem of land subsidence, approved the land for the golf course, a cycling path, a campground, and other architecturally unintensive facilities. In time, when the land becomes more stable, the area may come to more closely resemble Odaiba, the built-up nearby entertainment island, which was once also a dumping ground, Reclaimed Lot 13 (see Endō 2004).

FIGURE 14 Wakasu Golf Links is built on an artificial island of garbage in Tokyo Bay. Exhaust pipes all around the course disperse the highly flammable methane and other gases given off by the rotting waste.

At one time a Japanese institution like Wakasu Links might have been less forthcoming about its odoriferous foundations and polluted origins; yet Wakasu readily advertises its relationship to waste and sustainable practices. The course describes, on its homepage, that the island was built using "household waste, etc." *(katei kara no namagomi nado no haikibutsu)* from 1965 to 1974 and covered by 350,000 cubic meters of leftover construction dirt from Shinjuku, at a cost of 550 million yen. It goes on to say that the course processes the greywater it uses for irrigation in a facility powered by solar energy.[2] Japanese institutions, corporations, and other entities garb themselves selectively in the raiments of sustainability, just as institutions and corporations in other parts of the world do. Yet the ways in which sustainability is framed and marketed to the public, and the ways Japanese invoke and enact sustainability in their daily lives, reveal the contours of sustainability *Japanese*-style, also shedding light on environmental policy, environmental engagement, and attitudes toward ecology and environmentalism in Japan.

Previous chapters have outlined Japan's relatively serious waste predicament, and the Japanese state has instituted a massive recycling and conservation drive intended to limit waste. Yet in order to achieve this goal of waste reduction, government institutions have been forced to attempt to transform an extremely successful postwar campaign that turned frugal Japanese into consumption-oriented supporters of domestic industries. By the mid-1990s, Japanese bureaucrats and politicians

were attempting to balance waste curtailment and economic promotion—two aims that more often than not operate at cross-purposes. To
achieve their goals, they appropriated the notion of "sustainability," a term
that has engulfed Western environmental discourse, as a rallying cry.

In this chapter I examine the Western origins of the contemporary idea of sustainability, and then I evaluate the importance of outside
international pressure *(gaiatsu)* in Japan and the role it plays in shaping Japanese policymaking. Throughout, I balance portrayals of state
policy with glimpses of local or regional implementation. I look at the
business of recycling by scrutinizing a start-up company of one of my
more entrepreneurial informants. Of course, the sorting of recycled and
other waste creates tensions in some local areas, which find themselves
blighted with what they consider to be an unfair share of the nation's
waste burden, and this theme leads us back to the troubled waste transfer facility in Izawa. I examine how the ambitious waste strategies of the
state played out on the ground in Tokyo and how community-level Japanese responded to, and shaped, Japan's vision of a pragmatic "sustainability" adapted to Japanese needs and with a particular emphasis on
utilizing waste to produce energy in this self-proclaimed "small island
nation, poor in resources."

Sustainability

The term "sustainability" arose in policy circles after the UN's World
Commission on Environment and Development report in 1983, in
which the goal of "sustainable development" was explicitly raised
(United Nations General Assembly 1983; World Commission on Environment and Development 1987; see also United Nations General
Assembly 1987). Chaired by Prime Minister Gro Harlem Brundtland
of Norway, and thereafter known as the "Brundtland Commission," the
WCED issued a report entitled *Our Common Future* that profoundly
influenced the global environmentalist project and the notion of shared
responsibility for conservation of the planetary environment (World
Commission on Environment and Development 1987). The notion of
sustainable development gained greater currency as the sheer scale of
societies' interventions into surrounding ecosystems became clearer
and more widely publicized, and the imbalance between human settlements and their surrounding environments skewed more in the direction of desertification, pollution, dwindling stocks of wildlife, reports of
the greenhouse effect, and other perceived blights.

But "sustainable development" is a clear contradiction in terms. While small settlements with low populations relative to available ecological resources can be "sustainable," forms of technology available in most parts of the world now allow extreme human interventions into the ecology in strikingly short amounts of time.[3] Furthermore, many parts of the world, particularly in East Asia, are densely populated and require extensive appropriation of resources to support large urban populations.

The contradictions inherent in the phrase "sustainable development" have, however, by no means curtailed its usage—quite the contrary. Indeed, groups who emphasize sustainability coexist cheek-by-jowl with groups who emphasize development, and at times these parties are likely to think that they share more common ground than they actually do. In this sense the phrase operates very much like ambiguous symbolism in religion, social movements, and nationalism (e.g., the cross, the "peace" symbol, or the American flag; see Kertzer 1988); the very ambiguity of sustainable development allows more people to agree with it.

The notion of sustainability, then, permits compromise between environmentalists and industrialists by allowing the goals of the opposite camp to seem less threatening. People use the notion of sustainability to press for very different goals. For example, greens invoke sustainability to push for an agenda that is, in many ways, the antithesis of economic development, while corporations appropriate the term to give their activities a thin veneer of environmental sensitivity. There is a limit to how far each side will go, however. Environmentalists may rail against the detrimental effects that economic development has on planetary ecology, but many governments are unwilling to go so far as to condemn development when trying to balance commercial interests with more environmentally oriented concerns. This was the position in which the Japanese government found itself more or less beginning in the 1990s.

Japan's waste predicament was gradually becoming clearer during the late 1980s, when cities such as Tokyo began to grasp the urban implications of increased consumption that resulted from Japan's growing affluence. But this limited awareness did not immediately transform into serious policy changes because of Japan's continuing obsession with economic growth, particularly during the irrationally effervescent Bubble years (pace Shiller 2000). These priorities were reflected in the balance of power in the Japanese bureaucracy at this time. While colossal ministries—such as the Ministry of International Trade and Industry, the Ministry of Finance, the Ministry of Construction, and the Ministry of Transport—commanded large numbers of staff and considerable resources, the Environment Agency was a relatively understaffed agency under the auspices of the prime minister's office.[4] Indeed, due to the Liberal Democratic Party's postwar structural dependence on public works as a means of consolidating and placating their power base in rural areas in particular, Japan's preoccupation with construction initiatives has long been considerable (McCormack 1996, 2002; Amano 2001; Kerr 2001). In 2009 the Democratic Party of Japan resoundingly defeated the LDP, and in the debt-ridden aftermath, chances are fair that the destructive postwar logic of the "construction state" *(doken kokka)* will continue to undergo changes. However, in the 1990s, with politicians and bureaucrats firmly—indeed, structurally—oriented toward economic growth and development, it was unlikely that the Japanese state would take strides toward conservation and moderation without considerable encouragement.

The phenomenon of *gaiatsu,* or "outside pressure," is important with regard to the complex of forces that nudged the state along a more "sustainable" path. During the postwar period, Japanese politicians have responded to and occasionally depended on international opinion, even open lobbying and cajolery, to institute difficult policies (e.g., Pharr 1990, 191, 231). Politicians and other officials may save face by diverging from their domestically safe position only when international pressure builds to such an extent that a change in position seems reasonable to Japanese sensibilities. Because of the added luster and authority of the Western, particularly regarding environmental issues, Japan in the late 1980s began to respond to pressure for environmental restraint that might have been more difficult to achieve without such pressure. Yet other influences played an important role, and any changes effected through external pressure certainly occurred glacially. After dissemination of the Brundtland Commission report, among other influences, the currency and appeal of sustainable development began to grow. In particular, amid the international pressure that built up just prior to the 1992 United Nations Conference on Environment and Development (UNCED) in Rio de Janeiro, Japan found discursive cloth with which to swathe its need to curtail waste. In 1991, in preparation for Rio, Japan amended the nation's Waste Disposal and Public Cleansing Law to emphasize reuse and recycling (Japan Environment Agency 1997a, 4.A.1). The international declarations made at Rio had an eventual effect on Japanese politicians and citizens alike, with the rhetoric of sustainability gaining some currency. But the collapse of the Bubble economy also brought considerable pressure to bear on Japan's choices. Significantly, the idea of "sustainability" was made to mean different things in diverse contexts to suit the political ends of the state.

Competing Pressures

While the Japanese government does respond to *gaiatsu,* it is important to remember that postwar Japan has been quite capable of defying world opinion in favor of domestic prerogatives. The most obvious case involves Japan's attempts to finesse its wartime past and its resistance toward shrill Chinese and South Korean diplomatic pressure.[5] But given the subject matter of this book, I focus instead on the much-debated and controversial case of Japanese whaling.

Like other nations—the United States, Russia, Iceland, and Norway, to name a few—Japan has a long history of whaling, having embarked on the practice in the twelfth century (Kalland and Moeran 1992). With advances in whaling technology and on-ship processing, however, the balance between depredation and replenishment has shifted to such a degree that numerous species of whale have been deemed in danger of severe depletion or extinction, in the 1970s and 1980s in particular. Japan campaigned vigorously against efforts to curtail whaling led by Greenpeace's effective Save the Whales campaign, which had succeeded in helping transform the International Whaling Commission (founded in 1946) from a pro- to an antiwhaling entity. However, due to the relatively unambiguous evidence of species endangerment and the severity of international reactions to the reluctance of nations like Japan, the *Schedule of the International Convention for the Regulation of Whaling* was amended,[6] bringing about a de facto moratorium on whaling (International Whaling Commission 1986a, 1986b; Stoett 1997). Ever since the signing of the agreement, Japanese officials—not to mention whalers and Japan's apparently dwindling number of whale-meat enthusiasts—have bristled at the restrictions it imposed. In recent years Japan has challenged the extent to which a convention designed for the management of whaling has been employed by conservationists to abolish the practice of whaling (e.g., International Whaling Commission 1989, 159–164; Mayer 1997).

Japan and its prowhaling allies argue that the decisions of the IWC, including its Scientific Committee, are dictated as much by sentiment and political concerns as by firm "science." This is ironic, however, since the cetacean "science" in the name of which Japan hunts whales

is widely regarded as all but useless by non-Japanese experts (for a
scathing critique, see Clapham et al. 2006; pace Morishita 2006). Pro-
ponents of whaling point to the fact that the convention's moratorium
on whaling was supposed to last until the species populations returned
to less-diminished levels, but that the moratorium was prolonged by a
commission dominated until recently by abolitionists. (Japan has used
development funds in the Third World to encourage essentially non-
whaling nations to join the IWC and vote sympathetically with their
patron, along with other manipulations of the process.) Japan has also
challenged what it sees as unfair strictures of the convention by embark-
ing on (and then increasing the amount of) putative "scientific research"
on whale populations (Mayer 1997, 1–4, 5–11, passim). Between 1987
and 2004, for example, Japan culled 6,777 minke whales alone as part
of its Antarctic research program (International Whaling Commission
Scientific Committee 2006, 41), and its total "scientific research" culls
through 2007 reached 11,389 whales.[7] After these culls, Japan allows
the whale meat to be sold on the market, and I have for years found
whale sashimi available seasonably in Japanese stores. Critics contend
that Japan attempts to undermine the convention and keep its whaling
industry alive through this growing trickle of whale supply, as well as
allowing Japanese consumers to retain a taste for whale meat (e.g. *Science* 26 April 2007). However, the *Asahi Shimbun* reported in 2008 that
purveyors of whale meat were finding it difficult to sell the meat made
available each year, even after slashing the price, and have abandoned
the sector in response to longstanding international revulsion toward
Japanese whale hunting.[8]

Importantly, Japan's whaling establishment justifies its opposition
to the convention in terms of sustainability. They claim that the population of Antarctic minke whales, for example, has reached such heights
(perhaps in excess of 510,000, according to a rough 1989 estimate by
the IWC Scientific Committee, but currently under stringent and heated
review)[9] that careful whaling would be more than "sustainable" enough
to protect the populations. Furthermore, Japan claims, speciously, that
the protected populations of whale are so high that they have developed
out of balance with their ecology, rapaciously consuming species of fish
and thus depleting the catches of Japanese fishermen. Morishita Jōji,
then deputy director of far-seas fisheries at the Japanese Fishing Agency,
claimed that the whaling controversy "'raises very important principles
about the sustainable use of resources, and the role of international
law'" (quoted in *Observer* 24 June 2001, 11). Morishita, later represent-

ing Japan's delegation to the annual meeting of the IWC (23–27 June 2008) in Santiago, Chile, expanded on these remarks:

> Japan's objective is to *resume sustainable whaling* for abundant species under international control including *science*-based harvest quota and effective enforcement measures. At the same time we are *committed to conservation and the protection of endangered species.*...
>
> Decisions in the IWC should respect science, international law and cultural diversity. Consistent application of *science*-based policy and rule making together with *the principle of sustainable use* is the paradigm for the management of living resources accepted worldwide....
>
> Claims to the contrary are an emotional response without *scientific* foundation.... [T]hese countries do not have the right to impose their ethical or moral values on Japanese *as long as whales are sustainably utilized.*...
>
> Therefore Japan, together with other members supporting *the sustainable use of whale resources,* have expressed their commitment to normalizing the IWC.... This means protecting endangered and depleted species while allowing the *sustainable utilization* of abundant species under a controlled, transparent and *science*-based management regime. (Delegation of Japan 2008, 1, 6, 7–8 [emphasis mine])

Japanese officialdom here has clearly mastered the sly rhetorical art of invoking "science," "sustainability," and other jargon to serve their ends. It must be said that IWC proceedings are dominated by political theater, with both sides responsible at times for selective use of data to suit their purposes—some more than others. Japan is deeply culpable in this regard, with a whole whaling research infrastructure of frequently risible scientific merit (see Clapham et al. 2006). Western scientists suggest in response that Japan is actually collecting this data for future whale harvest on a much larger scale (e.g., *Science* 26 April 2007, 534). Japan's professed solution to an alleged fish scarcity is to bring whale populations down through "sustainable" culling in order to let the ecology recalibrate, and they have produced numerous "scientific" analyses to justify their case.[10] Japan's stubbornness in resisting outside pressure regarding whaling—an almost Gallic assertion of cultural exceptionalism—demonstrates that the nation does not simply bow to international opinion with reference to sustainability. It also shows that Japanese politicians are willing to use, and shape, the terms of sustainability to pursue their own agendas.

Japan's choices regarding sustainability have, of course, been influenced by the country's economic downturn. The Japanese state's reflex-like emphasis on pump-priming measures to restart economic growth—for example, through massive public works infrastructure spending—placed economic development squarely in opposition to "sustainable" concerns. Yet the situation was more complicated. There has been a clear pattern of frugality, conservation, and general restraint during downtimes in the Japanese economy. In contrast to much of the developed West, where environmental sensibilities grew more widespread partly due to affluence, in Japan belt-tightening measures have been more palatable in times of difficulty. Garon shows that amid the rubble of defeat immediately after the war and after the oil shocks and recession of 1973–1974, Japanese groups that urged austerity found willing ears in contrast to what happened during the exuberance of good times (Garon 1997, 192–193, passim). A similar sociocultural phenomenon began to take shape toward the end of the Bubble period, which has influenced contemporary sustainability consciousness in Japan in important ways.

The Sustainable Life

Until its assertive rise as an industrializing world power beginning in the latter half of the nineteenth century, Japan long stood as a model of how to achieve sustainability both at the village level and on a grand scale.

In early modern Japan, canny frugality was widespread, extending variously from necessity, social protocols, and Japanese notions of ethical behavior. Farmers and others in rural areas had long been kept at subsistence or near-subsistence levels due to punitive taxation of rice yields. Under this and other austere conditions, these Japanese were forced to avoid waste wherever possible (see Duus 1976; Sippel 1998; J. W. White 1988). Alongside this rural population living close to the blade, samurai elites themselves observed an ethic of frugality and self-denial in accordance with the code of the warrior. Tokugawa Ieyasu, the first shogun, sternly enforced this self-restraint: clause seven of his "Rules for the military house" *(Buke sho-hatto)* states that, "All samurai in all fiefs must live frugally" (Sansom 1964, 7–8).[11] Correspondingly, great emphasis was placed on strength of will, articulated in terms of the quasi-mystical Japanese notion of *ki* (spirit, or energy), which eventually diffused beyond the samurai caste. During World War II, Japanese state propaganda played upon this notion of self-discipline and sacrifice—as well as the innate strength and purity of the Japanese "race"—in order to squeeze maximum efficiency out of the general population and redirect resources toward the war effort. The idea of "spirit" pushed Japanese to extremes of self-denial and privation (e.g., Benedict 1947, 23–26). Furthermore, wartime Japanese society was firmly oriented against waste, with slogans such as "'extravagance is the enemy'" and "'waste not, want not, until we win'" (Havens 1978, 15, 49). Frugality is a consistent theme in both rural and urban community studies conducted from early in the Western anthropology of Japan into the postwar period (Embree 1939; Norbeck 1965; Dore 1973; R. Smith 1978). The Japanese state played a key role in encouraging thrift, both before and after the war, for all walks of life. In part with the cooperation of women's neighborhood organizations, Japan was able to instill parsimony at the deepest level, in households. Attempts to "rationalize daily life" and "'eliminate waste not only in food but in everything'" arose as a response to early postwar prosperity in the 1950s, and frugality began to be linked with economic nationalism—conserving resources would lead, according to this rationale, to lesser dependence on foreign powers (Garon 1997, 185–186, 192).

Significantly, this concern with thrift and waste went hand in hand with Japan's longstanding perceived scarcity of natural resources across the archipelago. The notion of Japan as a "small island nation, poor in resources," or *shigen shōkoku nippon,* among other expressions (Dinmore 2006), emerged as a powerful discourse in the early twentieth century whose stamp on Japanese society endures in the present day. Rising Japan coveted the abundant land and resources of its mainland neighbors, such as China, and seizure and exploitation of these resources was a major objective of Japan's colonialist project (Dinmore 2006; Young 1998; Sato 2007). Japan's discursive stance juxtaposed "scarce" domestic resources with a looming, expanding Japanese population. This, coupled with elites eager to engage in ambitious state-directed resource management projects, ensured that resource development in Japan was pursued as a means toward safeguarding national strength—with intermittently paranoid or peaceful, integrative stances (Dinmore 2006; see also O'Bryan 2000, 2009). Against this historical backdrop, the state's postwar selection of the term "resources" *(shigen)* to index recyclables such as metal cans was a culturally powerful, highly resonant choice that largely won over the general populace whose cooperation in separating recyclables from waste was being solicited. Japan's "poverty" with regard to resources is, of course, a matter of perspective; while Japan is neither a petrodollar-brandishing member of OPEC nor a resource-abundant North American nation, the "green archipelago" is hardly barren moorland (e.g., Totman 1989), and Japan's seas have been rich in fisheries until only very recently, due in part to careless, decades-long discharge of toxins into Japanese waters (e.g., Broadbent 1998). Suffice it to say that, in Japan, frugality and resource-conservation have long been framed as social virtues and as elements of national competitiveness.

This strong competitive bent left its mark on Japan's ecology. During the postwar decades of growth-at-any-cost, abundant waste plagued the Japanese archipelago in the form of toxic pollution both in hotspots where production occurred and more diffusely, in ambient form. Clearly, there existed forms of waste—such as slag heaps, contaminated air and water, and other environmental pollution—that were deemed more socioculturally acceptable, at least for a time, and, conversely, others that were not. The latter forms were targeted in the Herculean efforts of Japanese corporations to combat waste and inefficiency in production, encapsulated in a business philosophy known as *kaizen.* Japanese for "[continuous] improvement," or "betterment," *kaizen* was a highly influential disciplinary strategy of team organization in which every possible improvement in manufacturing or other processes would be aired, debated, and implemented as part of the very structure of the corporation. Believing that the project of countering waste is an endless series of small steps, rather than the result of occasional grand reform programs, Japanese multinationals such as car manufacturers and electronics giants are structurally oriented toward employee input, troubleshooting, and fine-tuning on a continual basis (Ohno 1988; Imai 1986). In the logic of high-speed growth, waste was framed in terms of resources and cost rather than byproducts and environmental defilement. But it is interesting to note that the seeds of responsible environmental behavior—if the state chooses to cultivate them—may lie in the very same focus on intensity, detail, transparency, control, and hard work that has served Japan's export-oriented industries so well.

Except for several decades during the postwar period, then, Japan has been a nation that has, generally speaking, shunned waste and lived relatively sustainably, though less so as time has gone by. Daily life generally revolved around use of organic materials, and extensive reuse spanned a broad range of spheres. Even human waste was husbanded conspicuously (and fragrantly) rather than divorced from everyday life.

Tokugawa Japan was characterized by a very open attitude toward acts of elimination. Fecal matter was kept separate from urine, and while both were exploited as fertilizer, excrement was highly prized as "night soil" (e.g., Watanabe Z. 1991). Receptacles were arranged along thorough-fares so that farmers could collect travelers' excrement, and Edward Morse writes that in the late nineteenth century, due to the importance of night soil, tenement houses actually *lowered* the rent charged on a room if there were a higher number of occupants, with no rent paid if there were five or more (Morse 1961, 231–233; Seidensticker 1991a).[12]

The rapidly growing capital had significant waste to handle. From at least 1655, waste disposal became wedded with land reclamation, as urbanites struggled to deal with the city's waste and to create—from marshland, bogland, and other unsuitable tracts—new land on which to expand. These comprise large areas of what is contemporary Tokyo.[13] Edo had, from this time, clearly designated disposal sites, though refuse found its way into most available areas (see Endō 2004, 789 ff.). Clearly, though, human waste in densely populated cities could become putrid, noxious, and foul if left to fester. Urban Japanese displayed unusual resourcefulness here. In Edo there existed an extensive symbiosis between the urban precincts and outlying agricultural areas, with sig-nificant organic waste processing in what became the world's most pop-ulous city by 1720 (Kobayashi 1983; Watanabe Z. 1991; Hoshino 2008; also Cybriwski 1998). Some historians argue that it was precisely urban Japan's success in controlling the miasmatic conditions of waste—par-ticularly waterborne disease spawned by human waste[14]—that led to Japan's relatively steep trajectory of development vis-à-vis mainland Asia during the early modern period. Farmers who came to Tokyo to sell their produce could use some of the proceeds to purchase human waste to bring back as fertilizer for their lands.[15] In a variety of forms, Tokyo—via its waste—diffused throughout much of the surrounding Japanese countryside in an early augury of the capital's present-day toxic wastescape. Some of these historical and geographical overlappings of waste and land-use are poignant. Rural Japanese long made sustainable use of areas of wooded commons outside villages called the *satoyama*. While the *satoyama* was integral to village life decades ago, many such villages—often depopulated through urban migration of younger gen-erations and as a result of other factors—now neglect the *satoyama* (e.g., Sprague 2005), making them ideal sites for illegal dumping of waste from Tokyo and elsewhere (Sprague, personal communication).

Apropos of Japanese whaling, it is interesting to note that some Japanese informants in Horiuchi framed the differences between Japanese and foreign whaling in terms of sustainability. For many of my informants, respectful Japanese consumption of whale was an example of sustainability and ethical treatment of animals from which other nations could learn. For example, in a local sushi restaurant my middle-aged friend Shin-chan voiced an impassioned defense of how, historically at least, Japanese whalers had treated the leviathan. While, according to Shin-chan, foreign whalers focused solely on whale oil and dumped the rest of the carcass, he emphasized how Japanese whalers used every part of the whale, therefore treating the animal, and the sea, with greater respect. "They even plucked the whiskers to use for musical instruments!" he cried, strumming an imaginary shamisen with his hands.[16] (Conversation then turned to the controversial practice of using a funnel or tube to force-feed geese to make the French delicacy foie gras.)[17] This kind of rhetoric over sustainability made whaling seem nobler, and Japanese practices more ethical, to Japanese informants who were weary of international criticism of Japan generally, from World War II to contemporary environmental policy.

During Japan's period of high-speed growth, frugality in daily life largely fell out of fashion, pushed aside by new habits of consumption (see Clammer 1997). The success with which Japanese achieved prosperity, amid much rhetoric regarding hard work, duty, and sacrifice, led to affluence that further increased consumption.[18] By the 1980s and early 1990s, conspicuous consumption had become commonplace, and young Japanese knew at best only secondhand the privation that their parents or grandparents had experienced. This trajectory of affluence and prodigality came up in informal narratives by Horiuchi residents of the growth years and the more recent downturn of the Japanese economy. The moral nuances of "waste" were articulated repeatedly in these collective narratives. Not infrequently older residents seized upon waste to critique younger Japanese, whose behavior often seemed the antithesis of traditional notions of reserve and sacrifice emblematic of Japanese group-mindedness. Yet there were unexpected points of convergence between the values of young and old Japanese. During meals at the Ōtsukas' home, for instance, Mrs. Ōtsuka and her husband would correct the manners of their two postadolescent sons, as parents do at family meals the world over, but they were less stern than when the boys were younger. Waste in both meal preparation and clearing leftovers was discouraged, and when Japanese friends joined them for a meal (for example a girlfriend who had gotten to know the family), they might feel obliged by their own upbringing and by the social situation to "perform" frugality, helping responsibly with cooking, serving, or saving the left-overs afterward. The boys, living under the same roof in Horiuchi and able to take part in meals, were expected to work for their own spending money. Yet when outside the normative space of the home, these young Japanese displayed their own ideas of waste. One of the Ōtsuka sons, Haruhiro, spent an entire bitter winter wearing only a sweater. At first, I thought he was merely trying to convey some sense of style or stoicism. But when I jokingly brought it up with his parents, Mrs. Ōtsuka said, chuckling, "He's stupid *(baka)*. He could buy a jacket if he wanted to, but he spends money on other things." Haru, unlike his mother, thought it was a waste to blow money on a jacket he could do without when he could spend a fraction of that amount on a date in trendy Odaiba. Some older informants were not so charitable. One woman in her early sixties, talking about *wakamono* (youth), equated lack of thrift with a lack of dis-cipline, lamenting: "Young people waste food," the waste of rice being a particularly glaring faux pas in Japan (see Ohnuki-Tierney 1993). "They don't have any discipline *(shitsuke)*," she added. "When I was younger, I had to kneel *seiza* next to the table with good manners. Nowadays, chil-dren can sit any old way. There's no backbone anymore."[19]

Despite this criticism, many of my younger informants were fairly frugal when they went to live out on their own. Such twenty-something Japanese in Horiuchi and nearby tended to save and reuse leftovers, set aside and fold (sometimes) the wrapping paper from gifts they opened, and show other signs of thriftiness that were both the product of their upbringing and not unrelated to their self-supporting circumstances. Knowledge of and participation in the Western-style environmental-ist discourse were also key differences with some of their elders. But it is important to note that such practices as the intensive, many-lay-ered wrapping of store purchases or gifts in Japan, seen by Westerners as wasteful and environmentally unaware, have long been considered important, even essential elements of courtesy and good breeding in Japan (see Hendry 1993). Nonetheless, there are signs that such ingrained practices are undergoing a sustained shift as public interest in the environment and sustainability in Japan increases.

Sustainable living has, indeed—slowly—come back into fashion in Japan, though its uptake has been by no means universal. Take one of Japan's successful retailers. In the 1980s, a time characterized by the most extraordinary excesses of conspicuous consumption, the supermarket Seiyū created a small line of lifestyle design products called Mujirushi

Ryōhin (now known outside Japan as "Muji"). Literally meaning "quality no-brand goods," Mujirushi encouraged the simple life, with sleek, no-nonsense products that harked back, in a contemporary manner, to the *wabi-sabi* minimalism and simplicity of Japanese aesthetics. The Mujirushi line as a whole was, significantly, a direct critique of wastefulness and brand-obsession pandemic at the time in Japanese society, notable for dispensing with excessive packaging and relying on word-of-mouth and a good catalogue over paid marketing (e.g., Watanabe Y. 2006; Ugumori 2006). Mujirushi's line of food products included no-frills versions of standard contemporary Japanese fare such as shiitake mushrooms or spaghetti. Mindful that the odd, broken bits and ends of sliced mushrooms were often discarded by retailers, Mujirushi reasoned that they could sell these scraps for less, and the scraps would taste the same. Similarly, spaghetti is cut to a uniform length, and the ends are typically disposed of. Mujirushi decided, however, to sell these ends separately as a bargain version of spaghetti, which also reduced waste in their food production cycle.

The simple values and humbler *raifusutairu* (lifestyle) that Mujirushi products evoked resonated with Japanese entering a period of belt-tightening.[20] The post-Bubble hangover, as it were, was also characterized by a sense of atonement. Informants voiced their impression that after the collapse Japanese society seemed to "wake up" to social and environmental problems that had remained underexamined during the boom. As informants like Andō-san commented, expressing the high-on-the-hog frenzy of the Bubble years, "Who had time to think about the environment back then?" This social upheaval, though uneven and gradual, created a discursive space in which rhetoric over sustainability could begin to influence Tokyo residents shaken by the end of the boom times, by evidence of insalubrity and pollution in their living environments, and by waste problems in the capital, including incineration and the blight of vermin crows. Some of this rhetoric surfaced in the mass media, but some also flowed from an active government campaign to construct sustainability that drew on Japanese cultural references and reflected state priorities with regard to waste and pollution.

Before detailing the evolution of "sustainable" waste policy since the late 1990s, I first want to describe the state of waste collection and recycling in Horiuchi when I arrived there in 1998. This concrete ethnographic material will help frame the specifics of government policy with regard to recycling, waste disposal, and sustainability and dramatize the considerable changes on the ground that had materialized in communities by 2009.

Recycling Horiuchi: Form over Content(s)

Household waste in Horiuchi was separated into different categories and dominated by Tokyo's dependence on incineration to free up landfill space. In addition to "burnable" and "unburnable" collections by the metropolitan government, much of the community's recycling had previously been undertaken informally by proactive members of a senior citizens group there, a form of voluntary collection known as *shūdan kaishū*. Some members of the group collected metal cans and newsprint every other Friday morning in my immediate neighborhood in order to raise money for group activities. Commercial establishments, on the other hand, had to sort out their recyclable waste themselves. Along the main commercial artery, restaurants, bars, karaoke establishments, and so on, had their recyclable refuse collected either by the suppliers, who exchanged empty bottles in plastic crates for those newly delivered, or by a private recycling company that would handle collection, cleaning, and sometimes reprocessing.

When I took up residence of Horiuchi, observance of waste protocols was an important feature in the responsibilities of every resident. Indeed, this was one reason why landlords throughout Tokyo were reluctant to take on foreign tenants—foreigners, they reasoned, were untutored in the minutiae of waste protocols and unconcerned with the ramifications of noncompliance, which could lead to trouble and local embarrassment for the landlord, as well as fines. There were, however, sometimes more racist motives for this exclusion (see Chapter 6). Because every resident belonged to a multihousehold waste cluster, responsibility was typically shared among four households, and great care was taken to maintain at least a veneer of propriety due to the rather public nature of regular waste collection.

The category of "unburnable" waste is revealing in this regard. Community shame can often rest on easily visible infractions or improprieties, so it is not surprising that residents focused overwhelmingly on visible manifestations of waste compliance. The officially recommended plastic bags for waste storage and collection were translucent but only partially transparent. Smaller households or single occupants like myself often decided to reuse the more or less equally opaque plastic bags left over from visits to the local supermarkets and other such establishments. Since the contents of these bags were not easy to verify from the exterior, the monitors for each waste cluster would rarely if ever take the trouble to inspect what was actually inside the bags at their small waste stations *(gomi shūsekijō)*. Far more emphasis was placed on having the bags out in time for collection, on positioning the bags correctly at the appointed waste station, and on making sure that the waste station was swept and tidy after collection took place.

Informants' responses in interview reflected this preoccupation with the container over the contained, or the form over the content(s). While my landlady took great care to instruct me in the proper separation *(bunbetsu)* of waste in my first days as a technical member of her household, many of my informants—particularly middle-aged and older men—confessed that they were less than painstaking when it came to making sure that the correct discards went into the stipulated bags. Mr. Iwasaki, my landlady's husband,[21] made a dismissive gesture when the topic was raised: using an informal word for wife *(kanai,* literally "inside the house"), he retorted, "[she] does [all that]," washing his hands of such unmanly concerns. This occasionally blasé attitude came out in the comments of others as well, but usually not until after one or two interviews. Satori, a young woman who lived on the outskirts of the community in a high-rise, had at first responded with characteristic earnestness about the importance of conscientious waste separation as dictated by the authorities—only to admit later that she could not be sure whether she always succeeded, for instance, when she came home late after a night on the town with friends. Yasuhiro, a teenage boy preparing hard for college entrance exams, was sheepishly aware that he should have been more engaged in waste-related matters: "There's no excuse *(mōshiwake nai desu ga),* but I'm always studying..." He revealed later that he left drink cans and other packaging around the house as often as he disposed of them himself and that he could not vouch for his care in separating waste.

There was a range of active compliance, then, with waste protocols. The more or less invisible condition of waste when bags or plastic bins were placed out at waste stations meant that it was all too easy for residents to feel that compliance in separating contents was more voluntary than compulsory. And at this juncture in the late 1990s, without strong convictions regarding the importance of recycling (or respecting the logic of the government policies in question), some residents found it easy to toss a drink can into the "unburnable" bin rather than to set it aside for separate recyclables collection. As time went on, though— through 2009—state measures and public awareness of environmental issues generally led to greater compliance.

It was precisely this need for instilling awareness of and cooperation with the state's recycling efforts—and statistical indications of mis-allocated waste at the receiving end (e.g., Matsutō and Tanaka 1993; Nakasugi 1985; Watanabe K. 2003)—that led to an intensified recycling campaign during my fieldwork. Yet pro-recycling and pro-sustainability policies had been "on the books" for a number of years, a feature in speeches and white papers but underimplemented and satisfying contradictory objectives.

The Evolution of Policy

After the Japanese government made key alterations to the Waste Disposal and Public Cleansing Law in 1991 in order to limit the waste burden, the state drafted the Basic Environmental Law (1993) and promulgated the Basic Environment Plan (1994). The last two documents in particular marked a profound shift in government policy and ensured that "sustainability" would become a mantra of the Japanese state. Nevertheless, these early documents display a predilection for bold pronouncements without the inconvenience of binding timetables or other forms of accountability. In fact, much of this early policy-writing leans heavily on the "categorical imperative" (that is, abundant use of the word "should") in articulating Japan's environmental future. These ambitious statements seem particularly unrealistic when they stray into areas of social behavior and consumer activity. For example, the Basic Environment Plan calls for citizens to "reduce the environmental burden caused by daily activities":

> They should refrain from using their private automobiles when such use is neither necessary nor urgent. They should also save electricity, reduce the amount of pollutants, such as detergent, they release, decrease total waste and cooperate in sorting waste for recycling purposes. (Japan Environment Agency 1994, III.3.1.4.B)

The proscription against unnecessary use of the automobile seems especially naïve, as many of my informants who owned cars possessed them as much for reasons of status as for "necessity" or "urgency." Though these dictates do, of course, reflect ambitions that any far-thinking government planner might share, when accompanied by nothing more than rather vague implementation strategies, these official commandments express a desire for waste reduction without any real expenditure of political capital.

In the mid-1990s Japan mobilized to honor commitments it had made at the Rio summit and to meet the challenge of hosting the December 1997 Conference of the Parties to the United Nations Framework Convention on Climate Change, held in Kyoto. That the rest of the world would be coming to Japan to negotiate sustainability and emphasize collective responsibility for the planet's ecology (what later came to be enshrined in "the Kyoto Protocol") created a strong incentive and one that became extremely powerful in transforming public opinion with reference to the "environment" *(kankyō)*. Informants described this period as a time when there would usually be at least one article every day regarding *kankyō* on the front page of each major newspaper. Politicians participated actively in this discourse on environment. For instance, Prime Minister Hashimoto Ryutarō made the following pronouncement at a special session of the United Nations General Assembly on 23 June 1997:

> Now, let us renew our determination and seriously consider concrete measures to promote sustainable development, a goal upon which we agreed at Rio de Janeiro.... From the depths of postwar devastation and despair, Japan has achieved rapid economic growth since the end of the Second World War, experiencing severe pollution problems in the process.... There is perhaps no other country that can share both the suffering of a developing country and the concerns of a developed country to the extent that Japan can. This is why Japan makes it a national policy to cooperate in the promotion of sustainable development. (Japan Environment Agency 1997b,3–7)

Yet while the government was willing to discuss the importance of sustainability and responsible ecological behavior in the abstract, particularly when giving speeches overseas, policy tracts from this period are characterized by a strong vein of economic rationalism that reveals the Japanese state's simultaneous preoccupation with the nation's post-Bubble economic revitalization (e.g., Japan Environment Agency 1997a, 4.A). Moreover, they dangled the prospect of profiting in the near future from the development of eco-friendly technologies as an appealing counterbalance to impending environmentalist curbs on economic activity.

These early documents described the larger project of sustainability as "building a society allowing sustainable development with a reduced environmental load" *(kankyō e no fuka no sukunai jizoku-teki hatten ga kanō na shakai no kōchiku*) (Japan Environment Agency 1994, article 4; see also Japan Environment Agency 1997a, 4.A). While sustainability is clearly an aim, this phrasing does make it clear that human burdens on the ecology, though reduced, are unlikely to be eradicated. We also in this period begin to encounter the term *junkan-gata shakai* (recycling-oriented society). Later, as I show below, the emergence of the concept of "zero-emissions" came to signal Japan's vision of a sustainable society where waste is recycled back into production, a concern of particular resonance in a resource-anxious nation such as Japan.

However, after the conflagration of toxic pollution anxieties in 1999 in particular, the tone of policy documents on the subject of waste reduction more generally changed abruptly and heralded more sustained attention to waste issues through 2009.

Dioxins and the Material Cycle

The outcry from the Tokorozawa dioxin scare certainly demonstrated the extent to which this insidious form of *kōgai*, or environmental damage, could ignite passions and morph into a politically destructive public cause. Politicians and bureaucrats made public a joint-ministerial and agency action plan less than two months after the flare-up of the nationwide scandal. The document, entitled Basic Guidelines of Japan for the Promotion of Measures against Dioxins, was approved on 30 March 1999 and was revised at the end of September of the same year. It sketches out a much more aggressive approach to Japan's waste dilemmas and produced effects that were visible in my fieldsites.

Lest there be doubt in anyone's mind as to the tense sociopolitical backdrop to this action plan, the Ministerial Council on Dioxin Policy Japan gets right to the heart of the matter on the first page: "Research on dioxins conducted by the Government has confirmed the safety of vegetables and tea, particularly in the area of Tokorozawa City in Saitama Prefecture. However, it is necessary to alleviate citizen concern...." (Japan Environment Agency 1999c, 1.3).

The action plan vowed a staggering 90 percent reduction of national emissions, at 1997 levels, by 2002. Many of the subsequent provisions are reminiscent of previous calls for recycling and waste reduction, but the sense of pressing need and the sharp tone reflected in these measures are striking. Assembled under the rubric "Urgent Measures," the Council sets a deadline for the establishment of "target volumes for waste reduction," calls for new systems of industrial waste disposal to be "quickly built," and calls for an "'industrial waste control manifest system'" to improve accountability for waste disposal (Japan Environment Agency 1999c, 2.1, 2.6.1, 2.6.2, 2.6.5). The action plan aims, furthermore, to meet a target of recycling 80 percent of construction-related waste. Incinerators in schools and other smaller institutions are banned (e.g., Japan Environment Agency 1999c, 2.6.8). Clearly, the government was taking a bolder stand on sustainability and health through waste policy.

Much of this post-Tokorozawa action plan makes specific reference to the sensitivities of public opinion and suggests strategies to assuage concerns. The Council calls for institutions to expand the publication and distribution of materials explaining the waste problem "in order to obtain the understanding and cooperation of the public regarding the dioxin issue" (Japan Environment Agency 1999c, 2.7.2). Furthermore,

> [E]ach Ministry and Agency is to strive to provide accurate information relating to dioxins, by providing information to citizens' information centers, and consumer information centers around the country, as well as through publications, the Internet, and the mass media, etc. Through these activities, seek to eradicate concern about dioxin pollution in food and mother's milk, etc. (Japan Environment Agency 1999c, 2.7.3)

These initiatives toward greater transparency in Japanese institutions and the new alacrity in setting concrete objectives in waste policy were remarkable. Yet implementation on the local level led to some patchy results.

Changes on the Ground

The transformation of waste policy in Horiuchi was noticeable and sustained. Early in the summer the Tokyo waste bureau and others distributed posters and fliers informing the public of a change in waste collection methods *("gomi no shūshū hōhō ga kawarimasu")* at the end of June 1999. Whereas, previously, burnable waste was collected thrice weekly and unburnable waste once weekly, now Tokyo would scale the burnable collection back to twice weekly in order to accommodate a recyclables collection once a week. This maneuver was, essentially, an example of sustainability Japanese-style: Tokyo could implement recycling without incurring the higher expenses of an extra, separate collection day. While most of my informants were amenable to the idea of a recyclables day, some were irritated that burnable waste collection had been reduced by a third. Mrs. Furuhashi complained, "What do we do with the waste? If it is not collected, the waste stays in the house. And with raw garbage *(namagomi)*, it begins to smell, especially in summer. It's a pain, but they just want to save money and don't think [about us], do they." Nevertheless, there was little resistance to promotion of sustainability, particularly since it was phrased in terms of waste and scarcity of national resources, themes familiar and persuasive to Japanese. The notion of "resources" reliably calls to mind an idea of sacrifice for the greater national good with which many Japanese are sympathetic.

The weekly "resources," or recyclables, collection consisted of paper, divided separately into bundles of newsprint, magazine paper, and cardboard, all tied with twine; glass bottles and glass jars, collected in a yellow plastic container; and metal cans and tins, collected in a blue plastic container. The classification of recyclables became extremely specific, particularly important in a society like Japan where, it can be said, god is seen to be very much in the details. The documents outlining recycling also serve as a kind of inventory of the Japanese household, an unintended benefit for the social scientist. Whereas books, leaflets, *manga* comics, and the like were considered permissible recyclable paper waste, rolled fax paper, photographs, and envelopes with plastic windows were not. Broken light bulbs were forbidden, as were spray canisters and drink cans that had been used as ashtrays. Reusable beer bottles had to be returned directly to stores, as did the clear plastic drink bottles that are commonly recycled into shoes, fleeces, and other everyday objects. These latter bottles were returned to Japan's ubiquitous convenience stores, even if the bottles had not been purchased there, and with no refund, as in some U.S. states.

This proliferation of newly classified objects (and for some, new activities related to their separation, preparation, and collection) brought about some noticeable changes in the behavior of residents, and, for a few members of the community, some confusion as to their proper implementation. Because these new regulations were mandated by the state and reinforced by existing local social waste networks, the need to recycle was accepted more or less without question by nearly everyone, albeit with some scattered griping about how plastic bottle disposal entailed greater burden or about the time-consuming work ushered in with each new set of regulations.[22]

Yet in some ways these recycling measures dovetailed with rhetoric surrounding recycling and sustainability that had already begun to pervade Japanese discourse since the early 1990s. Policies such as the use of recycled paper for government documents and brochures, as well as the increased use of eco-labels on products (such as *chikyū ni yasashī— midori wo mamoru,* or "gentle to the Earth—protect [our] greenery") contributed to the gradual infiltration of an unaccustomed sustainable paradigm. Mass-media coverage of the passion for environmental causes and recycling in Europe—Germany in particular—and in North America helped lay the groundwork for this intensification of policy in 1999 and lent a certain moral cachet to these causes. The Japanese state, in turn, passed laws that enforce recycling of appliances and other electronic goods by requiring manufacturers to recycle their own products after use. For example, the Home Appliance Recycling Law (2001) allows consumers to return used refrigerators, washing machines, wall-unit air conditioners, and televisions they discard to respective manufacturers for recycling, and a 2003 amendment to the Law Regarding the Promotion of Effective Utilization of Resources requires manufacturers to recycle personal computers returned to them by consumers.

The Profit of Recycling

Household-based recycling of waste was not the only domain where pragmatism reigned. Businesses, and those in the business of recycling, were overwhelmingly influenced by expediency. While sympathetic with the idea of sacrifice for the nation, entrepreneurs and business people were rarely if ever willing to operate at a loss to do so. Regulations, the threat of fines, and attractive financial incentives coaxed the commercial sector in the direction of greater cooperation with the task of waste reduction and recycling.

At first, Mrs. Ishimoto's business entailed the collection of glass bottles from area restaurants and drinking establishments. Beverage distributors paid Mrs. Ishimoto to gather the bottles at regular intervals, clean them, and sort out the ones that were broken or otherwise defective. Then she returned the bottles, saving the distributors this laborious (and odoriferous) work. Mrs. Ishimoto's employees also compressed collected metal cans and other recyclables, acting as specialist intermediaries who knew the community and the network of clients who used them as agents for their waste. Recycling businesses sometimes offered to buy recyclables such as metal cans and tins from commercial establishments, making money for themselves and saving these establishments the collection fees the state charged for nonhousehold waste producers.

Mrs. Ishimoto recalled that recycling really became an issue on the local level only in 1994, not long after the state began taking reduction and recycling of waste seriously. She found a market niche and exploited it, and her building became inundated with "resources" receptacles. When asked whether she had a garden, she exclaimed: "It's on the roof....everywhere [else], there are bottles!" During the Bubble years, she said, "no one was thinking about recycling....But after the collapse, everyone began to start thinking about it....It's because Japan doesn't have resources *(shigen)*...It'd be a waste to use things one time only." By summertime 1999, Mrs. Ishimoto suddenly found her business catapulted up to a new level of magnitude by political events. The capital wished to intensify its drive to recycle and needed private companies such as Mrs. Ishimoto's to supplement metropolitan efforts. With few companies in Tokyo possessing the necessary expertise, Mrs. Ishimoto soon experienced a dramatic leap in business.

A persistent theme that ran throughout conversations with Mrs. Ishimoto was that she was certainly no environmentalist. A no-nonsense, but kind and considerate woman, she had, over the years, managed to support a chronically ill husband and a son, more or less singlehandedly. The business of recycling was, for her, an opportunity for profit, plain and simple. So while the state's efforts to reeducate its citizenry with regard to sustainability and waste reduction will likely continue to bear fruit, for the industries springing up to address Japan's waste predicament, it will clearly be important for the state to continue to provide financial incentives to enlist the efforts of this unsentimental sector.

Fighting for Sustainability Locally

As is nearly always the case for large-scale processes such as the state's protean recycling campaign of the 1990s, national initiatives necessitated interventions into local milieux, and the burden of undesirable side effects tended to fall most heavily on such hapless local communities. Izawa's troubled waste facility served as a transfer station for recyclables as well as unburnable waste destined for landfill. Thousands of tons of waste passed through this transfer point on the way to its ultimate dumping grounds, with recyclables sorted for separate processing. Being a local problem rooted in a specific place, objections tended to be specific as well. Even aside from the emissions of dangerous toxins causing illness amongst Izawa residents, complaints ranged from the banal (odors coming from the facility; the clear increase in large truck traf-

fic and accompanying exhaust fumes) to the visceral (queasiness over
the enormous quantity of "unclean" waste from all over Tokyo flowing
through the community). Yet it was also clear that proximity to waste
sharpened residents' opinions about environmental issues generally,
with the toxic dispute lurking as the inevitable subtext to most discussion about health and environment.

Significantly, wider public anxiety surrounding dioxins created
a political climate in which the previously stalled Izawa dispute could
move forward and begin to garner attention from higher authorities in
the Tokyo Metropolitan Government. Through the summer of 1999, the
protesters' plight began to receive greater attention, particularly media
attention, due to the newfound resonance of toxic pollution in public
discourse surrounding health and environment. In the early autumn,
protesters found themselves arrayed against the Tokyo waste bureau in
an assembly room of the ward government offices in front of an audience of a special committee of ward assembly members. (Of course,
"arrayed" somewhat misrepresents the spatial politics of the hearing.)[23]
Here, the demonstrated mendacity of officials in Tokyo's waste bureau
received particularly scathing comment from some prominent ward
politicians in attendance. The case was then moved, at last, from the
ward to the Tokyo docket, the level at which the protesters' arguments
against the facility could finally be aired efficaciously. Tokyo was now in
the difficult position of needing to find a resolution to the convincing
case that the Izawa protesters had made. For the protesters' case was
difficult to ignore in this sensitive climate; yet due to the gravity of the
capital's waste quandary, Tokyo needed the waste compaction and other
benefits of Izawa's waste transfer facility more than ever.

The Tokyo government first agreed in early 2000 to pay the medical bills of residents who, according to physicians' reports, had suffered
from symptoms common to those of Izawa protesters and the category
of Azuma Disease that they had identified. Then, in mid-summer, the
waste bureau concluded that the entirety of the illness problem actually derived from toxic gas entering into the ward sewage system from
the facility. According to this official explanation, waste liquid from the
stored trash seeped down through drains in the facility and was collected in a tank below. Because the liquid in the cesspool was sometimes
not emptied for a week or two at a time, organic matter in the cesspool transformed into hydrogen sulfide, a noxious and soluble gas. The
bureau concluded that these fumes had entered the sewers, seeped out
of indoor household drains (for example, the open drains of a shower

or bath), and entered the air inhaled by residents. The Tokyo government finally declared that the facility had dealt with this chemical seepage once it had been discovered (the facility alleged it had stopped using the cesspool some months later), and on the basis of this explanation of community toxic exposure, Tokyo granted limited compensation to affected residents.[24]

After the dust settled and bureaucrats and others felt comfortable speaking candidly about the Izawa case, it was clear that there existed a gap between the ward's explanation of events and that of Tokyo. In 2009 I returned to the ward offices to interview members of the policy management department and the waste management section of Azuma Ward. When asked what had caused the flare-up in symptoms near the facility, they expressed their belief that the demolition and disposal of the state institute for machine technology formerly located there may well have caused unusually toxic chemicals to be released into the surrounding community. Since the institute had existed even before World War II—and therefore could, in theory, have contained wartime research materials of murky provenance in addition to other dangerous chemical agents—its vast number of stored chemicals (some of unknown composition, according to these ward employees) could have been emitted into the air or the sewage system, triggering sharp reactions in Azuma Disease sufferers that made them hypersensitive to any subsequent environmental pollutants in the area, even in very low concentrations. (This, indeed, mirrored some of my own private speculations.)

However, when I raised this version of events with a bureaucrat in the Tokyo waste management bureau, he objected vociferously. The Tokyo bureaucrat believed there was no evidence for such a conclusion, and he believed that the official explanation—of inadvertent hydrogen sulfide pollution seeping into sewage pipes—well encapsulated the scientific findings the metropolitan government had accumulated in its investigation. Over a decade later the Tokyo government remained adamantly opposed to the notion that the facility's construction or operations could have caused widespread misery in the community beyond the very limited admission of responsibility given by Tokyo in 2000.

The controversial events of those dozen or so years demonstrated that one result of toxic pollution in Izawa, whether officially recognized or not, was the ready adoption by protesters and their supporters of the terms of the environmentalist discourse, including those of sustainable development. Intriguingly, their embrace of the values of the global environmentalist movement was selective, with differences peculiar to the Japanese sociocultural context. GEAD member Kimura-san embodied some of these contradictions. Kimura-san worked as a playground monitor at a local primary school near the waste facility—indeed, a school shown by GEAD-commissioned testing to be exposed to high levels of pollutants. Amid the controversy surrounding toxic pollution in Izawa, Kimura-san transformed his role into that of a nature and ecology instructor for these children. Behind the school playground were slimy concrete pools filled with algae and plants, as well as a disheveled expanse of dirt, worn grass, and weeds. In this unlikely setting, Kimura-san conducted daily lessons on the microecology there with dozens of children. When I visited him, we discussed environmental issues while he helped several girls and boys find bugs and tadpoles with plastic cups and small nets. Many other children excitedly brought their finds to his attention. Kimura-san took great care identifying the specimens, describing what the creatures ate, how they survived, how they fit into the larger ecosystem, and so on. A conspicuous central lesson was the care with which these wildlife specimens had to be handled. In Kimura-san's outdoor classroom, what was taken out of these slimy pools and patches of dirt had to be returned unharmed. Thus the reason for Kimura-san's strategic choice of setting for the interview was clear: these children were put in danger by the toxic facility, and yet they also held the key to rebalancing Japan's future sustainability. Kimura-san mentioned children and ecologically responsible conduct repeatedly in our later discussion.

According to Kimura-san, there were particular challenges to sustainability and engagement in Japan: "The problem is, Japanese children think nature is dirty... [They] don't understand how fragile nature can be." He continued, "Adults, too. There's a balance *(baransu)* of ecology, an interconnectedness. Japan many years ago was different, but many Japanese today think everyday nature is troublesome, inconvenient." Despite his clear support for sustainable development, Kimura-san nevertheless voiced opinions on waste issues that bore an idiosyncratically Japanese character. Kimura-san was adamant that "we have to reduce unburnable waste.... We have to recycle much more," statements that were uncontroversial in global environmentalist discourse and which could impact, in theory at least, the future of the troubled waste facility in the community. But with regard to incineration of waste, he very much agreed with many of my other Japanese informants, who had little problem with increasing the incineration of waste as long as toxic emissions could be controlled: "Japan...has to burn its waste...the amount is increasing dramatically *(dondon fueteru)."* Kimura-san nevertheless cast considerable doubt on how trustworthy official assurances of incinerator safety would be. He referred to the GEAD's experiences with the Tokyo waste bureau during the Izawa controversy: "They take lots of measurements everywhere but only use the ones that are low.... They even did that with the school—they measured twice but, since one of the readings was above the limit, they only used the other one.... I'm telling you, Tokyo's waste bureau is terrible *(hidoi)."* Kimura-san's comments, and those of others, made it clear that the rationale for incineration had so permeated the Japanese social context that even in the midst of a toxic pollution protest, and against the backdrop of a major nationwide dioxin scare, protesters found the government's campaign to increase waste incineration relatively unproblematic. Needless to say, the categories of burnable and unburnable—terms set by the government, for interested reasons—went unchallenged.

Significantly, in 2007 the Tokyo government itself undermined this distinction between burnable and unburnable waste as Japan began to eye mass incineration of both categories of waste as an easy way out of its waste predicament. In 2009 this new waste policy led to the highly unexpected closure of the Izawa waste facility, though it was not due, at least publicly, to any new sympathy with protesters' arguments. There was a certain irony in the decision to start burning "unburnable" waste almost a decade after the Tokorozawa dioxin controversy; but it was also tragic, since the state had, during the intervening period, taken bold strides toward turning Japan into a model of sustainable policy strategy.

Sustainability in the New Millennium

The turn of the millennium saw a dramatic intensification of state concern with sustainability. The Japanese government's elevation of the Environment Agency to ministry status during a reshuffling of bureaucratic institutions in January 2001 demonstrated a continuing resolve to promote environmental concerns. The new ministry remained dwarfed by other, more development-oriented ministries, however, and therefore lacked the power, resources, or in some cases the jurisdiction, to make a real difference in resolving important issues.

The Japanese state—along with advocates, scholars, journalists, and others—also began to incant the mantra of "zero emissions." This expression is generally acknowledged to derive from the Zero Emissions Research Initiative set up in 1994 by the Tokyo-based United

Nations University and was intended to build on momentum toward sustainable development demonstrated at the 1992 Earth Summit in Rio de Janeiro. The zero emissions concept attempts to control emissions by ensuring that waste is, as much as possible, used as an input in other production processes—that is to say, any emissions from one process are captured and used in other production, whose emissions are captured and used productively in still other processes, and so on. This ostensibly aggressive change in many ways suited the Japanese social topography. In my experience Japanese politicians, university presidents, and even corporate leaders generally display a great affection for invoking lofty principles and abstract goals toward which they may or may not make progress. Take, for example, former Prime Minister Hashimoto's grand pronouncements on Japan's "national policy" of sustainable development to the United Nations General Assembly in 1997: just a year later the UN determined that Japan was responsible for 40 percent of the developed world's air emissions of dioxins, the world's most lethal man-made toxins (United Nations Environment Programme 1999), due to a de facto "national policy" of aggressive waste incineration. Japan seemed to be in no great hurry to alter this waste policy until the sordid facts became public knowledge, particularly domestically in the wake of the Tokorozawa dioxin scare. Yet many Japanese genuinely seem to take to the idea of 100 percent commitment to a goal that is inherent in the "zero emissions" concept. One need only look at the annual zealotry of the high school baseball phenomenon or the extreme dedication of some Japanese workers and the notorious linked phenomenon of *karōshi* (death by overwork). Just as Japan mobilized the nation to fight "total war" in the middle of the twentieth century and harnessed its energies toward high-speed economic growth in the postwar period, so "zero emissions" evokes a campaign to wage total war against damaging or wasteful emissions. It is a campaign that seems, gradually, to be gaining traction, though with biases peculiar to the Japanese social context.

Conserving Their Energy

The Japanese state has long been acutely concerned with energy issues. It is no surprise then that Japanese attitudes toward waste incorporate a strong vein of apprehension over "waste" of energy, which continues to shape Japan's response to the challenges of a world with arguably scarce, expensive fossil fuels. Significantly, the mundane, burdensome sphere of waste management furnishes somewhat novel possibilities to produce energy, which has become of great interest to the Japanese state.

The new millennium saw Japan tighten up elements of its waste cycle. The year 1999, fraught with dioxin controversies, brought the phasing out of small, poorly managed incinerators in favor of large, state-of-the-art ones. Early in this reformist period, Tokyo launched the extremely strict waste separation regimen described above, which explicitly forbade, for instance, the burning of chlorine-based plastics *(beniru)*, such as the thin, clear plastic wrapping on CDs and DVDs. These were categorized as unburnable waste. Tokyo also oversaw the construction of increased incineration capacity during the 1990s. While this was seen to be a good thing at the time, overexpansion combined with some success with waste reduction efforts—for example, through intensified recycling—meant that these large incinerators often operated below ideal capacity.[25] This became a problem for two reasons. First, incineration at very high heat (in excess of 800 degrees Fahrenheit, or approximately 427 degrees Celsius) more or less eliminates dioxins and other toxins. But if the furnace is allowed to cool—as happens when there is diminished waste to feed the flames—then toxins can be released. The lower waste volumes threatened the very logic of public health safety that led to promotion of these large incinerators. Next, and crucially, state-of-the-art incinerators in Japan are now built to generate electricity through their operations, and the incinerators sell this energy to offset capital investment and operations costs and to generate funds. Lower than expected waste incineration created shortfalls in this energy and revenue production.

It was against this backdrop that Tokyo began, in 2007, to expand testing of its newest incinerators with an eye toward burning plastics, for years a policy taboo, particularly in the wake of Tokorozawa. Ironically, Tokyo now needed *more* waste, and one convenient—though contradictory—place to find it was in its own "unburnable" waste collections (e.g., Shinjuku Ward Waste Office 2008). Hawking a new, improved filtration technology that allegedly controls emissions more than before, Tokyo determined that its incinerators could begin burning plastics safely at high temperatures. Activists criticized this policy as poorly tested and also misguided (e.g., *Japan Times* 3 November 2007), though opposition has achieved no real results thus far. Authorities have named this process "thermal recycling" *(sāmaru risaikuru)* due to the electricity generated from the added incineration. Incorporating the term "recycling" lends the operation the veneer of green respectability; since other developed countries use the term "thermal recovery," this wording seems yet another case of clever propaganda infiltrating the most basic level of Japanese environmental discourse. It seems likely that irresponsible household disposal may now increase, since one key rationale for careful separation and recycling—the safe processing of potentially toxic plastics—is now undermined. Naturally, the problem of the greenhouse gases produced by incineration has, throughout, been downplayed or avoided as much as possible in state discourse.

One unforeseen consequence of this shift in policy toward aggressively burning unburnable waste was that the need for waste transfer facilities went down considerably. In a bizarre anticlimax after over a decade of protest, Tokyo and Azuma Ward decided to cease operations in Izawa in 2009. What waste that still had to be compressed was redirected to another waste transfer facility in Western Tokyo.

Azuma Disease sufferers and their supporters were pleased—like Awaki-san, who lived nearby—but Tamura-san, who had been in bad shape, was recovering and still so weak that she could not leave her home. Others, like Kamida-san, remained wary and were still trying to get over problems like the lingering damage to her respiratory system. As for Tsubō-san, she reported slow but continued progress from her refuge outside Tokyo, though she still required a nurse. The waste facility's closure registered, of course, but she was now concerned with larger environmental questions. She and fellow GEAD member Yamada-san collaborated in setting up an activist group dedicated to measuring environmental quality and battling against volatile organic compounds (VOCs) around Tokyo and elsewhere in Japan.[26] Part of the guarded reaction from some in Izawa derives from the continued uncertainty as to how the empty transfer facility will be used.

Japan continues to grapple with its waste burden. With an enormous number of recyclable PET bottles (used for mineral water and soft drinks) being collected all over Japan, new ways have been found to use this material in ways that diverge from public rhetoric over sustainability. Polyethelene terephthalate is commonly recycled into products such as fleeces, athletic shoes, and other everyday synthetic items. But in a nation where resources are especially valued when utilized for energy, plastic pellets made from recycled bottles are now used in steel mill furnaces as a cheap source of fuel (Ōhashi 2004), and the Tokyo government supports businesses that exploit plastic waste to generate power (Yokoyama 2007, chap. 3). Here, energy trumps sustainability.

Ever mindful of the archipelago's resource scarcity, the Japanese state began in the new millennium to move toward extracting energy from landfill gases. The most important of these, methane, is a vexing greenhouse gas—approximately 20 times more potent than carbon dioxide—produced partly through decomposition of organic matter. A great deal of methane is generated through the belching and flatulence of cattle and other creatures, and it is also released from the earth's crust during warming, for instance from melting permafrost in Siberia and elsewhere. Importantly, abundant methane is given off by rotting waste in landfill. Mostly since methane is a relatively clean-burning source of energy, Japanese bureaucrats began in 2005 to devise methods for capturing it during waste decomposition (Ministry of Agriculture, Forestry and Fisheries 2006). This is one example of how the economic and development-focused orientation of the Japanese state with regard to scarce resources can become an asset in the promotion of environmental objectives. The effort is, at present, mostly limited to test projects. But the principle is simple: collect organic waste from households and elsewhere, process it in dedicated facilities, capture the methane that is released, and use the gas to power the facility and create a surplus that can be used, for example, to supply power to surrounding communities (Japan Institute of Energy 2002). Less well developed at present is the question of how to capture methane from existing land reclamation projects like the Wakasu golf course (e.g., Yokoyama 2007). However, where energy resources are involved, we can now be reasonably sure that the Japanese state will find a way to harness underexploited troves of BTUs.

Japan has, indeed, begun a root-and-branch review of the possibilities of exploiting biomass energy potential more broadly. Called the Biomass Nippon Integrated Strategy (Ministry of Agriculture, Forestry and Fisheries 2006), the review envisions utilization of biomass at key stages of industrial production (for example, with paper mill waste, which is rich in hydrocarbons). Power plants that generate energy from biomass waste produced 2,520,000 kiloliters of crude oil equivalent in 2005, and the target for 2010 is more than double that figure at 5,860,000 kiloliters. Waste products used as heat, for example, in factories, produced another 1,490,000 kiloliters of energy in 2005, also projected to double by 2010. Separately, use of black liquor and waste wood—both byproducts of paper production—as heat sources in paper mills reached 4,720,000 kiloliters in 2005 (ANRE 2008). This willingness to "dig deep" for advantages in energy management is mirrored in the longstanding efforts of Japanese bureaucrats to reduce the inefficient use of energy in the daily lives of the Japanese. Reckoning that increased efficiencies could conserve energy resources, Japan has pressured domestic manufacturers to produce more energy-efficient appliances and other products. Unlike some nations that use minimum energy performance standards for certain categories of products, the Agency for Natural Resources and Energy's effective Top Runner program selects the highest efficiency product in each of eighteen categories (e.g., cars, TVs, refrigerators, air conditioners) and then makes that product the standard for all similar products to attain. Within four to eight years, each manufacturer must ensure that the weighted average efficiency of all its products in a given category is at least as high as the chosen benchmark product (Nordqvist 2006; ANRE 2009, 31). While the ANRE wields no clear punitive sanction, if a manufacturer fails to meet the agreed standard, this fact appears prominently on mandated labels (ANRE 2008, 20), conveying their relative inefficiency to consumers. Importantly, the annual energy cost of using each product is displayed clearly on the label, enabling shoppers to compare products' energy performance in yen. Such standards have led to impressive gains in household appliances, for example, with air conditioners becoming 40 percent more efficient relative to 1995 and refrigerators saving 78 percent of their energy expenditure annually, compared to 1992 (ANRE 2008, 22). Under the banner of increased "sustainability," then, the state is conserving energy resources for the nation and promoting more environmentally friendly technologies at the same time. The Japanese state also claims it has made significant progress with regard to specific waste-related initiatives—for example, following the Ministerial Council's recommendations on dioxin policy, the relevant ministries reported that Japan reduced national dioxin emissions, at 1997 levels, by a dramatic 96 percent in 2006 (Ministry of Environment 2007a, 1–3; see also Ministry of Environment 2007b) and had "recycled" 92.2 percent of the nation's construction-related waste (Ministry of Land, Infrastructure, Transport, and Tourism 2006, 1; see also Ministry of Land, Infrastructure, Transport, and Tourism 2005, 23).[27]

Considering these recent advances, it can no longer be said that Japan is "behind" *(osoi)* in terms of sustainable development, though there remains characteristic Japanese emphasis on the *developmental* half of the phrase. Japan, with Tokyo at the fore in many cases, is attempting to rein in its waste problems, exploit opportunities to produce energy, and shape the environmental behavior of its citizens. Nevertheless, these efforts to influence Japan's waste shadow must still unfold over the sociocultural terrain of Japanese society, where attitudes to garbage, health, purity, pollution, and excess impact environmental engagement in profound ways.

Conclusions

The idea that one nation's environmental situation can be measured and compared with another rests upon the assumption that environmental "data" from one part of the world can be identified and made compatible with those from another. Certainly, this is true in some cases. Yet environmental engagement in complex societies is an intricate, nuanced phenomenon and resists such easy comparison. This book, for one, demonstrates that the "environment" in Japan, while offering numerous facets that shed light on other parts of the world, nevertheless remains intensely Japanese. In order to understand what environment in Japan really *means*, we must probe very deeply into strata of Japanese society that may at times be difficult to access and controversially at odds with the image that Japan wants to present to the world. Such data, furthermore, usually end up being murky, ambiguous, "thick," and resistant to numerical tabulation and other quantitative methods preferred by some analysts. While the spill of Japan's waste shadow over the archipelago and beyond can be surveyed by technoscientific measurements in an imperfect manner, and superficially analyzed using statistics, such an approach to waste issues neglects a crucial, highly resonant subjective dimension. As socioculturally moored as an anthropological approach may normally be, this volume's findings do nevertheless indicate that intensive ethnographic scrutiny of life and environment in one society can yield insights that improve understanding of socioenvironmental topographies elsewhere and impact international environmental policy.

There remains a certain tension between the specific and the general in anthropology and other social sciences, and this is notably so in environmental accounts, including the present volume. Many environ-

mental problems or health threats, particularly toxic pollution problems, have local origins, whether they be pollutant facilities releasing
toxins into surrounding communities (Bhopal) or tainted goods sent
elsewhere. Yet environmental problems also accumulate, modulate,
and drift across boundaries, even as solutions depend on provincial and
national governments whose jurisdiction remains fixed. Most, if not all,
environment-focused scholars are committed to environmental issues
generally—even the skeptics—and they tend to conduct their more cir-
cumscribed research with an eye toward broader processes and compar-
isons. In this vein, I myself address—albeit guardedly—current trends in
Japanese environmental policy and their potential impact.

Attitudes toward waste in Japan gave rise to the most visceral and
emotional of responses, including revulsion against "germs" or toxins in
the home, discomfort with interloping vermin, and disgust with regard
to perceived "untouchables" in Japanese communities—though, significantly, some Japanese I interviewed cared little or not at all about such
matters. Not only does this mean that efforts to interpret environmental issues in Japan must take account of such material, but any serious
attempt to understand contemporary Japan more broadly must also
engage with the complex ways in which Japanese think about waste, pollution, and linked exclusionary processes, which lie near the very core of
Japanese ways of reckoning. However, my research indicates that contemporary aversion to traditional "outcasts," at least, is not as virulent as
reported in the past. Moreover, concern with the "pollution" of death at
funerals seems to be on the wane (e.g., Suzuki 2000). I suspect that many
Japanese parents will bitterly resist their daughters or sons marrying
putative descendants of the customary "untouchable" outcaste for some
time to come. But judging by the rough, eventual adoption of the terms
of the global environmental discourse by sufferers and their allies in the
community toxic pollution protest I studied (see also Kirby 2004), for
example, as well as gradual discernible shifts in environmental priorities
over a period of approximately a dozen years, it is clear that attitudes
can change. Increased exposure to recycling initiatives and conservation measures, including the recycling of water in municipal swimming
pools, the spread of trendy secondhand clothing, scrap-sorting protocols for biomass energy initiatives, and so on, may all help diminish the
squeamishness some Japanese report with regard to waste-handling, to
the potential "germs" of strangers (yosomono), and to their very percep-
tion of what is pollutant.

International mass-media stereotypes characterize Japanese as

profligate consumers and polluters who are heedless of "the environment." While some elements of this portrayal may ring true, *Troubled
Natures* challenges such two-dimensional caricatures by attempting to
place contemporary Japanese rhetoric and practice in proper perspective. I have attempted to show that urban Japanese attend to, and are in
many ways deeply concerned with, their surroundings and participate
in elaborate discourses on "nature" that inform Japanese engagement
with their environs. Threats to reproduction, conceptions of hygiene,
notions of health and illness, shifting boundaries of place and the home,
exclusionary mappings of the community, evocations of nature—all
these elements commingled to influence the social climate of my fieldsites. Certainly, the ethnographic portraits presented in this book show,
at the very least, that there exists no stereotypical Japanese automaton
that accords with conventional international notions of how conformist Japanese people behave. On the contrary, my informants engaged in
varied and, in some cases, profoundly idiosyncratic relations with their
milieux that challenged any reductivist portrayals of Japanese society.

A rigorous analysis of environmental attitudes should cast its net
broadly to draw together the wide range of influences on environmental
thinking in any complex setting. Take, for example, the entanglement of
toxic anxieties and economic gloom that gripped Tokyo during much
of my research. Trepidation over dangers to community health, threats
to reproduction, and perceived risks to sperm quality, even "manhood,"
mingled with articulations of financial uncertainty, job-related fears,
and economic dread. Apparently distinct threats became more closely
interwoven due to their hooded nature; whether the specific cause of
anxiety was toxin-laden air, contaminated vegetables, the fall of the Nikkei stock exchange index, or the rising unemployment rate, informants
often found themselves subject to invisible forces over which they felt
they had no control. The fact that many also regularly observed murders
of crows ripping into waste stations in their community certainly did
not lift their spirits, either. Informants constantly transposed politics,
economy, community, family, and environment in their responses in a
way that underlined the importance of the broad perspective this book
has taken.

Clearly, societies engage differently with their surroundings. As a
result of distinct attitudes to "nature" and other powerful sociohistorical discourses, it can even be said that societies *see* their worlds differently. However, such generalized pronouncements, while sometimes
helpful, create the conditions for assuming that *all* members of a society

should accord with the generalizations in question—an obvious interpretive problem, and a common one. Take Japan. Due in part to long-standing Japanese ideas that derive from that society's religious tenets about the mountains and the sea being the abode of *kami* (spirits) (see Berque 1997a), Japanese appear to regard remote "wild" spaces with relatively more discomfort than, say, Germans or Americans might. Present-day Japanese who engage with elements of the global environmentalist discourse are probably less likely to feel this way than their great-grandparents, but such sociocultural anchor-points do still matter. Partly as a result, many Japanese still prefer, broadly speaking, to engage with "nature" through semidirected tourism, sampling local culinary delicacies and perhaps immersing themselves in a hot spring *(onsen)* in a stereotypical small mountain village nestled in the countryside (see Robertson 1991, 1998; Ivy 1995) rather than venturing into remote natural preserves with fewer amenities.

As important as such an observation about Japanese engagement with "nature" may be, though, it runs the risk of contributing to and exacerbating prevalent warped notions of Japanese distinctiveness that influence how both foreigners and native Japanese perceive Japanese society (see Chapter 4). Compare this analysis with Japan's long history of elite concern (even obsession) with resource scarcity on the archipelago, discussed in Chapter 8. Partly as a result, most Japanese respond very favorably to the government's branding of recyclables as "resources," and this helps boost compliance with recycling initiatives there. This sort of general conclusion about Japanese society and its environmental attitudes addresses societywide issues but manages to avoid the reifying quality of broad, uncontextualized pronouncements on Japanese "natures." While the anthropology of Japan involves balancing both sorts of broader commentary, I have tried to allow extrapolation from my specific findings to wider commentary on Japan while avoiding as much as possible discussion that devolves into hackneyed conclusions or stereotype.

Such an approach involves an awareness that Japanese society and culture are not "unique," as sometimes assumed by observers—and in ways that may hit close to home. My Japanese informants, for example, again and again voiced concern over purity and pollution, and this culturally resonant fixation on purity has influenced government policy in different periods of Japan's history. Before environmentalists seize on this tendency, however, they ought to confront the clear discourse on purity that pervades global environmentalism itself.[1] Environmentalists often put an extremely high priority on purity—take the Tokyo branch of a global environmentalist organization I approached, whose representative informed me that they had already given up on Japan's "tarnished" ecology but used the East Asian nation as a base from which to protect relatively untainted Siberian forest[2]—and this bias toward purity can, in turn, influence these activists' priorities. There is, in environmentalist activism, a clear slant toward zones touched as little as possible by human contact, effectively ranking parts of the world on a scale from pristine to defiled. For some environmentalists, it is as if such "untouched" oases are endowed with a kind of sacred quality. This established tendency reflects, of course, clear and measurable technoscientific evidence of adverse human effects on environments, and I do not dispute this. But "environment" as commonly invoked in contemporary activist discourse tends to conceive of planetary ecology as though humans are or should be edited out of the equation entirely. Such reflex-like, holier-than-thou, even—as it were—environmental-fundamentalist attitudes can get in the way of activists working *with* communities that dwell in or border such troves of environmental purity (see Chapin 2004; Brockington 2002; Nygren 1999; see also E. A. Smith and Wishnie 2000). The very idea of pure nature, unspoilt by human endeavors, leads amateur environmentalists away from a clear understanding of the profound ways that human activities have long influenced the ecology of remote environmental precincts that can seem like the essence of purity.[3]

It is tempting, of course, to try to use insights into Japanese sociocultural dispositions or ideological discourses—either from this book or from elsewhere—to influence policy enacted by the Japanese state. Japan's concern with resource issues—most prominently with reference to questions of energy production—could of course be exploited to shape how Japan, or major cities like Tokyo, might adopt more sustainable practices. Yet such high-level priorities can manifest themselves in unusual and unpredictable ways. For example, Japan's desire to extract combustible greenhouse gases like methane from decomposing garbage and other sources of waste on a massive scale seems an extremely convenient marriage of state concerns and environmental prerogatives, one that could be exported to other nations as well. But Japan's efforts to produce energy by incinerating municipal waste seem problematic if such a practice becomes an end in itself. (Reducing waste and encouraging reuse are, of course, comparatively more efficient and have less impact in macrosocial terms, not to mention the planetary effects of greenhouse gas emissions.) Furthermore, the contemporary practice of using plastic from mineral water bottles and the like as a source of fuel in heavy industry seems, from an environmentalist perspective at least, like a far-less-desirable strategy than stronger emphasis on reuse, recycling, energy conservation measures, and development of renewable energy sources.

Mid-1990s rhetoric in Japan and other nations held out the promise that technologies developed in the green sector could prove an important and lucrative export. Yet it is only recently that economies like Japan's have begun to appear more realistically positioned to begin doing so. Based on the findings of this volume, however, it is difficult to predict whether Japan will be expanding and exporting the largely positive, renewable technologies or the ones that exacerbate some environmental problems while solving others. Indeed, it is important to remember that facilities such as methane-capturing facilities and high-tech incinerators are not only out there in Japanese society, but have, at the same time, been spawned *by* that same system. The technology that a nation builds to deal with waste or energy can take a cultural form that may be difficult to transplant to foreign soil. On the national and transnational scale, for example, other nations may radically disagree with Japan's present rather cavalier attitude to greenhouse gas emissions related to waste management.

In the face of such complexity, environmentalism itself is certainly not seamless or monolithic despite prodigious efforts to make it so. Top-down global environmentalist organizations attempt to cover "the globe" but frequently fail to address adequately the myriad cultures that both complicate and potentially facilitate environmental activism. In the irregular patchwork of environmentalist nongovernmental organizations (ENGOs) and other local and regional, governmental, and transnational entities that comprise global environmentalism, there is already some attention to the sociocultural dimensions of environmental problems. But due in part to the distorting pressures of large-scale fundraising (Chapin 2004; Igoe 2003) and the distrust of local peoples in protecting resources and biodiversity (e.g., Fairhead and Leach 1996; Igoe 2005; see also Escobar 1998)—as well as the widespread preference for "hard" science and quantifiable metrics for judging policy success—difference is not given the priority it should be. Context-driven, socioculturally attuned environmentalism would be well served by anthropology's long-standing focus on the microlevel of everyday experience and its ability to engage with larger-scale forces and connections. Though it has a ways to go in this regard, environmental anthropology could continue to provide subtle mappings of important sociocultural topographies, helping to supplement or even direct some environmental policy through its findings. Rather than attempting to create a sprawling top-down environmental apparatus of massive scale fundraising and subsequent massive waste of revenues and opportunities (albeit with successes along the way), perhaps a better model would be that of the *wiki*—acephalous, adaptive, and empowering people and alternative sources of knowledge rather than attempting to exclude and control. This volume could then be considered a small, early step in that collaborative direction.

Perhaps for the environmental anthropologist in particular, it is difficult to resist the impulse to peer into the future and speculate as to how conditions and trends in the recent past might unfurl in the years to come. This I will largely avoid here. Japan's environmental policy is still in many ways an open book. But with the participation of a recently more proactive Japanese state in questions of waste and environmental health, and with the relatively strong influence of the state on conduct in Japanese communities, one can see how it might happen that Japan could become—in the not-so-distant future—a capital of recycling and biomass energy management. Furthermore, efforts to craft "sustainable" behavior in Japanese communities—recent developments in incineration policy aside—could continue to bear fruit. In this, the seeds of grassroots environmental renewal—whether or not the state chooses to help cultivate them—may lie in Japanese society's same focus on intensity, detail, control, industriousness, and teamwork that has served its export-oriented industries so well. Indeed, perhaps there is room here for an "environmental fundamentalism" that is distinctly Japanese, where forms of sustainability can succeed as much due to preoccupation with economic or energy imperatives as to any homegrown eco-conservationist sentiment. If so, such processes of sustainability and the forms of engagement to which they might give rise must take into account the deeply embedded notions of waste, pollution, nature, excess, and thrift presented in this volume if they are to take hold.

Notes

Chapter 1: Introduction

1. The chemical family commonly referred to as "dioxins" includes the planet's most lethal man-made substances: toxins that are not only carcinogenic but also attack the endocrine system, the immune system, and the reproductive system on a genetic level (Institute of Medicine of the National Academies 2003; World Health Organization 1999; Institute for Global Environmental Studies 1999). In 1998, at the start of my fieldwork for this project, Japan had by far the developed world's highest air emissions of dioxins due to its choice of waste management strategies (United Nations Environment Programme 1999).

2. Dauvergne's invocation of the shadow metaphor in his study of Japanese influences on deforestation in Southeast Asia is important but differs considerably from my usage. This is partly the result of his borrowing of the phrase "shadow ecology" from other work—e.g., MacNeill, Winsemius, and Yakushiji 1991, 58–59; Dauvergne 1997, 10—that balances the ecological capital that one nation takes from a second nation with the ecological shadow for which the former nation is responsible. Nevertheless, Dauvergne's is a different study—from the discipline of international relations—and has different aims.

3. Of course, few are willing to trade the conveniences and the fruits of development for an abstract ideal of "environment," a point to which I return in later chapters.

4. The term "environment," as commonly used in various languages, often evokes an idealized notion of "nature," or ecology, separate from human contact. I attempt to use the term "environment" so that it acknowledges the sociocultural interplay that environment has with ideas of nature, particularly in the Japanese context.

5. An analysis of Japanese terms for "nature" appears in Chapter 4.

6. "Nature" has stood as a much-contested term in anthropology for decades. Interrogated aggressively in feminist anthropological debates in par-

ticular (MacCormack and Strathern 1980; Ortner 1974; Reiter 1975; Ortner and Whitehead 1981; Delaney and Yanagisako 1995), though prominent in a range of anthropological discussions, nature comprises a social category often fundamentally opposed to culture. Yet nature is riddled with culture, not least in the cultural interpretation of the natural realm. Suffice it to say that I do not use the term lightly here as a simple description of surroundings.

7. Azuma Ward and the names of my two fieldsites there are pseudonyms necessitated by the terms of my interviews. I elaborate on this in Chapter 2.

8. While much of the methane in the atmosphere is emitted by live-stock—85.63 million tons annually worldwide through digestive processes and as much as 17.5 million tons more from the manure produced (United Nations Food and Agriculture Organization 2006)—much also emanates from landfill sites and other waste concentrations. To profit from this wastage, the Japanese government in 2005 announced plans to extract energy from food scraps, sludge, and other raw waste, mostly by exploiting the methane gas this waste gives off as it rots (Ministry of Agriculture, Forestry, and Fisheries 2006).

9. The moral dimension of condemnations of waste obscures, for Baudrillard, the fact that the very value of affluence depends on and is maintained through excess in social milieux controlled by scarcity.

10. Curiously, both Rathje and Murphy (2001) and Scanlan (2005) choose a relatively foreign term to describe their object of analysis. Scanlan, a Brit, dedicates his study to "garbage," even though this term is far less common than "rubbish" and other less printable terms in usage in the Queen's English. Cf. Hawkins' article "Shit in Public" (2004) for a bracing Antipodean variant. On the other side of the Atlantic, Rathje and Murphy, North Americans, choose the clearly Anglophile (or at least antiquated) title *Rubbish!* despite its infrequent usage in American speech. I detail their other, stated reasons below.

11. Mary Ellen Brown deftly sums up this complex of meanings: "First, trash connotes that which ought to be discarded, a sort of instant garbage; second, it connotes cheapness, shoddiness, the overflow of the capitalist commodity system. Third, it connotes a superficial glitter designed to appeal to those whose tastes are ill-formed according to the dominant perspective.... Fourth, trash is excessive: it has more vulgarity, more tastelessness, more offensiveness than is necessary for its function as a cheap commodity" (Brown 1989 [cited in Bird 1992, 107]). While Brown is referring to soap operas and other cultural forms viewed as "low" by elites, her description can of course be applied more broadly.

12. Using Rathje and Murphy's (2001) definitions as a reference point, Melosi appears to make "rubbish" synonymous with "trash" (as opposed to "wet" organic waste); see his common phrasing "garbage and rubbish" (e.g., Melosi 2005, 227), which he places under the encompassing rubric of "refuse."

13. For instance, he invokes the term "garbage" when it surfaced in historical discourse, such as when he describes the American concern of the 1880s

over the "garbage nuisance" (Melosi 2005, 2) and the furor over the "garbage crisis" a century later (195). But he is careful to define garbage as "organic waste" and uses the term exclusively as such where it surfaces in the historical record (e.g., 19). In addition, Melosi is sometimes content to use "garbage" as a general term in the American context, as are Rathje and Murphy, due to the word's currency in American speech.

14. There were interesting historical variations, however, as Melosi (2005) enumerates in his introduction. Ancient Jewish codes mandated remarkably prescient household waste and hygienic practices and in time the streets of Jerusalem were washed daily. The ancient Greeks became so perplexed by waste problems that they created the first known municipal dumps, and Athens was, apparently, the first government to rule against throwing waste into the streets. A medieval Paris wallowing in waste created a regulation whereby whoever brought a load of building material into the city had to cart out a load of mud or refuse. Despite these periodic innovations, however, waste remained a rank, putrescent blight on urban settlement from which humans were rarely (if ever) free.

15. Indeed, this all but inescapable stench of urban waste, rather under-emphasized in historical scholarship on cities, must surely have been a strong contributing factor in various societies to the rise of urban idealization of the countryside—in poetry, prose, painting, and architecture and garden design (cf. Classen 1993; Howes 1991; Williams 1973; cf. Süskind 1985).

16. In *Effluent America* (2001), Melosi writes that the city-based horse relieved itself of approximately twenty pounds of manure and several gallons of urine daily, a heavy metropolitan burden; Brooklyn alone had 26,000 horses at this time. Of course, the body of the horse eventually became waste incarnate: records show, for example, that New York City scavengers removed 15,000 horse carcasses in 1880.

17. Japanese habitually separated urine and feces when relieving themselves, and these resources, regarded as precious, were transported sometimes far out of the city for use as fertilizer (Morse 1961; Embree 1946; Macfarlane 1997; Sprague, Goto, and Moriyama 2000).

18. Jean Baudrillard (1998) writes that, when calculating gross domestic product, France and other countries glorify waste—pollution control, waste-processing, industrial discharges, etc.—in the (positive) production column rather than on the other (negative) side of the ledger, where it arguably belongs with other costs. Indeed, the division of economies into such binary terms is itself symptomatic of this capitalist logic.

19. This is particularly the case for those based exclusively in rural areas, whose proximity to and familiarity with the land usually does not offer fertile soil for such grand idealizations (see Knight 1997).

20. Note that the term "waste" in English derives from *vastus* in Latin, origin as well of the contemporary term "vast" (OED 1989).

21. Such waterborne diseases plagued most urban concentrations of the day, in Asia as in Europe (Macfarlane 1997 cites a range of relevant historical sources).

22. One conspicuous contemporary example of this logic of production is Odaiba, a trendy entertainment island development in Tokyo Bay. It was built upon layers of waste that then became prime real estate with expansive views of Tokyo's uninspiring bayscape.

Chapter 2: Perils of Proximity

1. The name is derived from the ward in which the facility was located. Azuma Ward and Izawa are pseudonyms, as are most of the names in this chapter.

2. Though some of the several hundred people with whom I discussed sociocultural/environmental questions in both my fieldsites were happy to go on the record, many definitely were not. I have therefore taken pains to alter the names of people and places, key biographical details, and some other elements in an effort to obscure the identities of my informants while retaining as much as possible of the ethnographic richness of my fieldsites.

3. The following sketch is based on informants' recollections. Throughout the rest of the chapter I excerpt quotations from field interviews (conducted between August 1998 and March 2000 and between September 2005 and April 2006, as well as during nine shorter trips, the last in summer 2009) and from questionnaires compiled by the Izawa-based protest group.

4. The full name of this office during the protest was the Tokyo Public Cleansing Bureau.

5. From most of Izawa's Forest Park the ventilation tower, which stands some nine meters tall at the top of a small rise, is the only visible sign of the subterranean waste facility buried below. Somewhat reminiscent of the architecture of I. M. Pei, with two matching pyramidal skylights embedded nearby in the grass and hedges of the hillock, the tower's rendering uses transparent plastic panels and an angular crown design to give these elements the veneer of high-tech. These protrusions at the same time appear autonomous, as if they were not appendages of the facility at all but, rather, mere sculptural features of the park.

6. More than a few also pointed out that *some* community had to take the facility. This stereotypically group-focused Japanese response is becoming less common in recent decades (cf. Lesbirel 1998; Aldrich 2008) and has to be understood in the context of a divisive protest-in-progress. Yet those who did not sympathize with the protesters sometimes characterized them not only as hypochondriac but as selfish as well.

7. Contemporary Japanese development projects, whether public or private, are full of such Orwellian euphemisms as "Forest Park," names that index

greenery or wildlife or breathtaking vistas in the relative (or, frequently, total) absence of the feature in question.

8. In spite of a decades-long stream of high-profile corruption cases involving politicians, high-ranking executives, and other leaders, respect for public officials remained, by and large, fairly strong in Japan, and government institutions retained high prestige. There was, for example, still fierce competition among top graduates to work in government bureaucracy. Yet by the time I had begun my fieldwork, a pervasive cynicism had taken hold in Japan with regard to the machinery of government and the structural failings and corruption of the establishment, and therefore trust in politicians and civil servants of all levels had undergone something of a sea change. By 2009 voters wrested away the long-standing conservative grip on power in repudiation of the LDP's post-Bubble stewardship in particular.

9. *Itai* means "painful" in Japanese.

10. The new laws had no teeth. They neglected to set specific standards or ceilings, left enforcement up to national and prefectural bureaucracies, and made no provision for funding regulatory enforcement. As Broadbent (1998) points out, these were typical of Japanese law, their ambiguity putting great power into ministerial hands. But they were a start.

11. While many in Izawa and elsewhere in Japan spoke of *taishitsu* without hesitation, some of the GEAD's hard-core members and supporters dismissed such folk conceptions as "unscientific."

12. Indeed, some years later, at the Japanese university where I worked, I was chatting in the elevator with my faculty's dean—a genteel, Kyoto-educated scholar—and I mentioned that I didn't often get sick. He joked that I would now have to start telling people that I was sickly, since I was a professor and had to keep up appearances.

13. Being a professional scientist, Arasaki-sensei was very careful about the claims he made. Indeed, in interview, he went out of his way to disassociate himself from the GEAD's wider arguments, since he hadn't compiled the data: "I can't vouch for all the things [the protesters] are saying about Azuma Disease.... But I *can* say that the Tokyo waste bureau twisted my data completely when they used it in their report."

Chapter 3: Mediated Anxieties

1. Ward authorities fought tooth and nail against this appropriation of their good name, though with characteristic Japanese reserve that sometimes bordered on the comical. For example, on the day of an important ward hearing, Azuma Ward staffers in the lobby of Azuma's ward headquarters *(kuyakusho)* stubbornly referred to the hearing only as the "Izawa Forest Park" case. When arriving journalists, photographers, and cameramen looked puzzled and asked where the Azuma Disease hearing was being held,

the firm reply from a female employee stationed by the elevators was that she did not know anything about Azuma Disease, but the Izawa Forest Park case was being reviewed on the fifth floor. Later, during the hearing, the mayor of Azuma Ward openly bemoaned that the name of the ward had become sullied through association with pollution and disease and urged all parties to resolve the situation without delay.

2. The Ginza is arguably Tokyo's most exclusive shopping district. The expression *sanpai-ginza* is, however, difficult to translate exactly. For example, it echoes phrases such as the "Nuclear Power Ginza" *(genpatsu ginza)*, referring to an area with a high density of nuclear power plants, and the "Typhoon Ginza" *(taifū ginza)*, an area of Japan where typhoons frequently pass. Put simply, the expression suggests that a place is the "number one" in Japan for a certain negative phenomenon—of particularly wry, ambivalent resonance in a hierarchy-obsessed society.

3. The narrative of this news synopsis flows from a close archival study of Japanese newspapers (the *Asahi, Yomiuri, Mainichi,* and *Sankei Shimbun* and the *Japan Times*), as well as some television news analysis, from early February 1999 into the summer. Particularly early on in the media saga, fewer sources appeared in the *Asahi Shimbun,* which was part of the same company as TV-Asahi and had a clear conflict of interest. There are too many individual sources to cite reasonably here. For example, taking one major newspaper on a single day, there were six articles on the Tokorozawa crisis, plus a front-page story on environmental hormones and an update on a major incinerator accident in Osaka involving high-level dioxin blood poisoning (*Yomiuri Shimbun* 10 February 1999a–h). Despite some minor variation between different news organs at times, the basic pattern of the coverage is consistent with this synopsis.

4. Since tea is usually steeped rather than directly ingested, the dioxin level was viewed as far less dangerous.

5. As it turns out, the only victims of significant toxic exposure there appear to have been employees of the facility in question (see, e.g., *Yomiuri Shimbun* 27 March 1999; *Asahi Shimbun* 26 March 1999; *Asahi Shimbun* 27 March 1999; *Yomiuri Shimbun* 30 March 1999).

6. Horiuchi was socioeconomically difficult to classify conclusively. The community was more solidly middle-class or above than Izawa. But both areas boasted landowners who were well off and professionals who lived cheek-by-jowl with more modest homeowners and struggling tenants. Azuma Ward is located on the more monied side of Tokyo from the "downtown" *(shitamachi)* neighborhoods where some other anthropologists have done their fieldwork (Bestor 1989; Kondo 1990; cf. Dore 1973). Since the population was also a mix of long-established families and newcomers, as with Robertson's fieldsite in Kodaira (1991), both communities were diverse in terms of residents' occupations and life histories.

7. In this way Tokyo—in essence an agglomeration of such communities

squeezed between and enveloping more built-up commercial districts—bears a greater similarity to London than to, say, New York City.

8. Some residents, particularly those employed elsewhere, treasured Horiuchi's ethos because the area had largely been spared development and the inevitable influx of homogeneous chain stores.

9. The storied peak is visible on these clear days either from hills in the capital (some of whose names still convey their prestige at having once been reliable viewing points) or from high-rise apartment blocks or other tall buildings with unobstructed views of the sacred mountain.

10. It is informative to note that, in the 1880s, before Japan's own industrial revolution reached its stride, Edward Morse (1961, 2) wrote: "The cities have an atmosphere of remarkable clearness and purity; so clear, indeed, is the atmosphere that one may look over the city and see distinctly revealed the minuter details of the landscape beyond. The great sun-obscuring canopy of smoke and fumes that forever shroud some of our great cities is a feature happily unknown in Japan."

11. Specialists, however, do conduct laboratory animal testing and analyze human case studies of environmental calamities and their aftermath, such as the 1976 Seveso disaster in Italy (see Roman and Pedersen 1998; Bertazzi et al. 1998).

12. Reportedly, there have been happily married couples who, after several months in a new home, saw their relationships spiral downward into divorce, or children who were focused, model students then experiencing sharp declines (e.g., Inoue 2004; Architectural Institute of Japan 2002), though I knew of no such cases in Horiuchi. As with Azuma Disease in my Izawa fieldsite, such afflictions can be easily explained away and causes can be exceedingly difficult to pin down.

13. A criticality incident in nuclear fission refers to an accidental chain reaction.

14. It was, however, by no means the first or the last grave lapse in management of Japan's nuclear power industry, a sector that has long been plagued by serious problems and secretive, incompetent management. Consider Tokaimura alone: the area hosted Japan's first experimental reactor, the Japan Power Demonstration Reactor (online in 1963) and Japan's first and second commercial reactors (1966 and 1978, respectively). Opposition to these plants was systematically played down at their inception, and Tokaimura's cooperation brought the community extensive state investment in infrastructure and services (Aldrich 2008). Yet problems surfaced repeatedly, such as the "loss" of seventy kilograms of plutonium dust in a plant there (reported in 1994) and a subsequent explosion and fire of drums of nuclear waste in 1997 that released radioactive fallout into the surrounding environment (Kerr 2001). News of disturbing lapses in management and maintenance dribbled out over the next decade, with serious environmental pollution incidents, casualties, and fatali-

ties at a range of nuclear facilities (see Problems with Nuclear Deterioration Research Society 2008).

15. In a debate with a friend about the developing incident, I seeded the conversation with a (truthful) comment to the effect that I did not know whether to trust the government's assurances that Tokyo was in no danger. In response, Sachiko-san, owner of a local café, commented, "I think that's a very good way to be thinking, Peter-san." The conversation then turned to the Japanese state's notoriously slow response to the Great Kobe Earthquake of 17 January 1995.

Chapter 4: The Cult(ures) of Japanese Nature

Epigraph: Cited in Moeran 1985, 254.

1. It is common in Japan to participate in *hanami* with co-workers, members of associations to which one belongs, and other groups. Therefore, a fair number of these Horiuchi residents were either pacing themselves for the coming evening or recovering from excesses of the night(s) before.

2. Though laden with irony, the phrase and the behavior it describes remain consistent with Japanese self-imaginings. While Japanese are, according to *nihonjinron* discourse, supposed to revere nature, they are also supposed to be group-focused and socially oriented (see Befu 1993, 109–13). People who used the phrase were poking fun at themselves, but also commenting on how *Japanese* the Japanese were.

3. Hoffman (1986, 19–20, 39) uses this image exclusively to describe the self-reflexivity of Japanese *tanka* poets, but I believe it merits wider application to engagement with nature in Japan.

4. For example, a Japanese expression denoting a person's strong aesthetic sense *(mono no aware)* implies an appreciation for the fleeting nature of beauty.

5. True to its namesake, Sakura Bank eventually succumbed to economic pressures in 2001, saddled with bad loans from the Bubble years, and is now known as Sumitomo Mitsui Banking Corporation.

6. For instance, I heard it used to refer to a person committing suicide.

7. It is important to note, however, that while many Japanese may at times display an affection for *nihonjinron* tenets, Japanese academia as a whole encompasses a wide range of serious scholarly pursuits.

8. Of course, any characterization of the fickle, mercurial Japanese climate as benevolent (e.g., Matsuhara 1964) seems clearly abstracted from the meteorological harshness of various seasons in the archipelago, and when this notion is coupled with claims of the Japanese having a peace-loving history, it loses any semblance of credibility, particularly among Asian societies with experience of twentieth-century Japanese aggression.

9. Yoshino's informants conceded that Koreans and Chinese, who share phenotypical features but do not have "Japanese blood," could "become Japanese" unless signs of their foreignness persisted in names, use of language, and

other sociocultural cues (Yoshino 1992, 118–119). Informally, these sentiments were echoed by my own Japanese interlocutors, though several insisted that it would "depend on the person."

10. Compare this with the "mixed nation theory" *(kongō minzokuron)* that developed in the context of Japan's imperial ambitions (Askew 2002, 79; Askew 2004; Oguma 2002).

11. Some of my more bookish informants even recommended I read Watsuji's work, since they recognized the affinities of his theories with some of the themes I was tracking in interviews.

12. For example, when you ask Japanese where they are from, you say, *"Furusato ga doko desu ka,"* or "Where is your *furusato?"*

13. I thank David Askew for his advice in translating part of this passage.

14. Such tree cover, among the few consistent sources of thick greenery in Tokyo, can be called a *chinju no mori,* or "sacred wood" (Berque 1997a, 89). See Berque (1997a, 88–90) for the affective significance in Japan of the "glossy-leaved forest" *(shōyōju-rin)* milieu, said to be that of prehistoric Jōmon culture, the alleged cradle of Shintoism.

15. Though the rice paddies that used to span the community were gone, the practice in itself was no more anachronistic than harvest festivals in other societies, including Thanksgiving in the United States.

16. Placards colorfully emblazoned with the names of products were written almost exclusively in the phonetic *hiragana* Japanese script. While an argument can be made for this making them easier for children to read, *hiragana* spell out *kun-yomi* pronunciations of indigenous Japanese words (*yamato-kotoba,* literally "words of the Yamato [folk]") without the need to use Chinese characters, so the practice carries a subtle but distinctive nuance of autochthonous chauvinist sentiment, particularly in the context of Shinto rituals.

17. Lighted red lanterns hung from the eaves outside an establishment symbolize a festive, alcohol-oriented, usually relatively cheap bar atmosphere. Chikurin conveyed this cheery ambience through its ornamentation but was slightly more upscale in the quality of its cuisine, the sophistication of its proprietress, and ultimately in price as well.

18. Their real homes could often be sources of great stress, as is true, of course, in other societies. For these patrons, Chikurin managed to satisfy, to some extent, the *idea* of home.

19. In a further sign of how important fellowship was to the *akachōchin,* Mama-san always kept a traditional-style sign hanging on the door outside (reading *shitaku-chū,* or "still in preparation"). This indicated to strangers that the establishment was not yet open. Since regulars knew to ignore the sign, this helped preserve the social intimacy of the place.

20. While it may not be the case for some delicacies, others such as vegetables and fruit are available most of the year from hothouse cultivators or foreign markets. Informants acknowledged that these complicated the idea of "in

season," but most staunchly insisted that these foods still tasted better when they were *shun*.

21. Berque (1997a, 17–38) authoritatively discusses the proliferation of nature-focused terms in Japanese and the potential insights to be gained by extrapolating from this terminological abundance. While it is important to know that there are more than twenty relatively commonly used words each for both "rain" and "clouds," more than five thousand words associated with the seasons, and many more thousands of *haiku* poems making use of them, I still find his subsequent explanation of the contemporary significance of this abundance less convincing (Berque 1997a, 43–44).

Chapter 5: Tokyo's Vermin Menace

1. One powerful tool in community Japan is the *kairanban*, a bundle of circulating notices that every household must read and to which they affix their seal, showing acknowledgement. The *kairanban* will commonly point out, in general terms, cases of neglected social duty, such as improper waste handling and waste separation or households that were slack in sweeping up leaves in front of their homes. Friends indicated that if a household resisted the polite dictates of the *kairanban*, they would probably face forms of ostracism or bullying *(ijime)* until the problem was resolved. One unusually well-travelled Japanese friend of mine was so sick of being singled out in the *kairanban* in her community that when she found out residents of fancy apartment buildings *(manshon)* did not have to read and sign the notices, she cried to her husband, "Oh, I want to live in a *manshon!*"

2. I asked Mrs. Iwasaki to be allowed to monitor the waste station for "our" household's shift, but it seems I was too foreign (that is, unpredictable) for such a sensitive social position.

3. The "bulky husbands" to whom I spoke made it clear that many of them did not like to be engaged in the very public, female-dominated, and rather demeaning activity of waste station monitoring when their household's turn came.

4. For numerous informants, touching the mesh with one's hands was a necessary evil akin to the quasi-polluted lid of a plastic bin—sometimes avoided by wearing gloves or holding antibacterial wipes.

5. Waste terms include *gomi* and *kuzu* (trash), *haishutsu* (emissions, discharge), *haiki* (waste exhaust), *haikibutsu* (a more scientific term for material waste), and *rōhaibutsu* (byproducts, such as in the body). Terms that reflect the moral or ethical dimension of waste, however, diverge: *mottainai*, which has religious associations (see Nickum et al. 2003), is used to mean "wasteful," "more than one deserves," or "not [being] worthy"; *muda* describes a pointless, useless waste of time, effort, or resources (e.g., *muda-dzukai*, "to waste money on").

6. Japan has a number of traditional corvine superstitions, such as, "'When

the crow calls, someone will die,'" or "'If the crows call in the evening, something unusual will happen'" (see Okuyama 2003, 199). These have atrophied a great deal in the present day with the ubiquity of the birds, but crows continue to retain such negative associations in Japanese culture, generally speaking.

7. This is not to say that Japan's woodlands are reliably salubrious. Pressures on wildlife in Japan's uneven wastescape include toxin-laden air that reaches wooded areas (see Chapter 3), toxic rural landfill, illegal dumping, problems of leaching, and other water issues.

8. In the name of interspecies understanding through cohabitation, Chim↑Pom, a Japanese artist collaboration, staged a 2008 gallery exhibition in Tokyo in which a man, a crow, and a rat spent three weeks together in a 3 x 2 x 2 meter enclosure, living exclusively off bags of garbage delivered daily from the Shibuya entertainment district, where the rat was allegedly captured. But the group was hardly an animal rights lobby: Chim↑Pom had provoked Tokyo crows in a separate work earlier that year, broadcasting a distressed crow's cry over a loudspeaker while on a scooter. The call filled the skies with a huge, unruly murder of angry *karasu* through several areas of the capital, a spectacle filmed in a work entitled "Black of Death" (*Japan Times* 14 August 2008).

9. For instance, the prototype of a trap used to catch *karasu* was created in Hokkaidō in the 1980s (Okuyama 2003).

10. E.g., http://www2.kankyo.metro.tokyo.jp/sizen/karasu/ (accessed October 2009).

11. Indeed, there was something of a precedent for this, as wartime cafeterias had served roasted sparrow to students as part of a government campaign to control scavengers of rice crops (Havens 1978, 139), but despite Ishihara's media-ready exhortations, crow pie in Japan has remained virtually nonexistent.

Chapter 6: Pure Obsession

Epigraph: M. Hasegawa 2006, 76.

1. I am indebted to Prof. Marilyn Strathern for focusing my attention further on purity with her comments on a paper I delivered in the Friday Senior Seminar (which she chaired), Department of Social Anthropology, University of Cambridge, on 7 March 2003.

2. While *burakumin* may seem like a monolithic category in mainstream Japanese contemporary usage, Ohnuki-Tierney (1987, 99) identifies the numerous itinerant and other groups (whom she calls "special status people") who were long ago lumped into the untouchable category due, in part, to their conspicuous difference from the largely agrarian, sedentarist Japanese social norm then.

3. Even the most tangential link to the *buraku* has been enough to condemn by association, as John Davis (2000) explains, citing a 1998 scandal that revealed the depth of the problem. Code words identifying an applicant as an "untouch-

able" had been written on a job applicant's CV by a professional investigator, despite the fact that the candidate was not in fact a *burakumin* or resident in a *buraku* community; he was merely resident near a *buraku* branch office located in a completely different area. Normally, such "guilt" by geographical association would have kept the applicant out of the job without the applicant knowing the reason.

4. "3-K" is Japanese shorthand for "difficult, dirty, and dangerous" *(kitsui, kitanai, kiken)*, referring to jobs eschewed by most Japanese but which allow others to find work—including foreigners and low-status or down-on-their-luck Japanese willing to incur such risks.

5. Day-laboring *burakumin* and other contemporary low-status workers have also apparently been systematically sought out as cheap, untrained labor to clean and repair equipment in the high-level radiation areas of nuclear reactors (Tanaka 1988; Horie 1984; Gill 2000).

6. This at least from a distance, in dramatic photographs, woodblock prints, or when seen through the haze from Tokyo itself.

7. This was hardly the extent of their precautions, however: according to Namihira (1977, 1984), hunters would traditionally refrain from sexual intercourse for a week before the hunt. They would also avoid using taboo words, such as words for polluted states or objects, and they would perform rites frequently, particularly to repurify their party if taboo words *were* used.

8. Indeed, it is not difficult to imagine clandestine *buraku* investors, having successfully infiltrated non-*buraku* Japanese society, also avoiding such real estate investments themselves for established financial reasons alone.

9. The campaign to make sumo an Olympic sport has been in an arm-lock for years due to the requirement of gender equality mandated by Olympic authorities.

10. Menstruating women in such societies can be prohibited from contact with hunting paraphernalia, from cooking meat, or from coming into contact with sunlight (e.g., Lévi-Strauss 1968); at the same time, women in some societies also become extremely powerful when having their period (Buckley and Gottlieb 1988).

11. *Uchi,* a term for "inside" or "[our] home" also enjoys a wider social usage, invoked metaphorically to index "our" group or "our" corporation in contrast to another. This usage is highly relational, as well as plastic, so that while the structural boundaries of the home are relatively fixed, ideas of social inclusion and exclusion can be adapted to one's emergent networks (see Kondo 1990).

12. I even saw him wear a mask to a funeral in the community, presumably for the same reasons as always, but the solemnity of the occasion dissuaded me from asking.

13. Traditionally, guests at a funeral will not re-enter the home until they have had salt sprinkled over themselves for purification after the polluting ritual

(e.g., Ohnuki-Tierney 1984). Contemporary funerals in Tokyo still commonly distribute packets of salt to guests for this purpose, and most informants said that they used them sometimes (though not always) or knew people who did. The practice is likely on the wane, however (see Suzuki 2000, 92ff).

14. While there is a great deal of interpretation and improvisation within these codes of hygienic practice—like whether one steps onto the "polluted" *genkan* floor when putting on footwear to insulate oneself from that plane of pollution—it is striking how enduring and pervasive these conceptions of hygiene have remained since the (albeit mythologized) days of raised wooden *geta* clogs and houses on stilts (see Pezeu-Massabuau 1981).

15. Reader (1991, 5–9) demonstrates that "belief" in the Western sense has always been low in Japan. Indeed, Japanese observance of religious practice and enactments of rites are more often a blend of tradition, routine, and social relations (Nelson 1996), in contrast to fervent, zealous belief characteristic of new religious cults (e.g., Reader 2000; see also Castells 2004).

16. For a detailed account of such rites, see Jeremy and Robinson 1989.

17. Ironically, some middle-aged ladies who lived on the lane were concerned about the hulking foreigner wandering along the tiny quasi-public alleyways between and behind the homes there—until I introduced myself and established my academic credentials with an official-seeming name card and some convincing questions in Japanese.

18. I thank David Askew for his insight into this social phenomenon.

19. Intriguingly, one informant, Miyamoto-san, used the term *kegare* to describe how a serious crime committed by a family member, such as murder, could plague a Japanese family like the mark of Cain *(kegare ha ichizoku ni kabutte-shimau)*. He was, however, an extremely well-read artist who had been exposed to Japanese folklore analysis.

20. The three monkeys are traditionally thought to represent wisdom and propriety, but Mr. Awaki invoked them in a clearly negative sense.

21. *Burakumin* and other contemporary outcasts apparently comprise high percentages of *yakuza*, or Japanese mafia, membership (e.g., Herbert 2000), and *yakuza* are heavily involved in illegal dumping, illegal trade in recyclables, and other waste-related activities (see Tanaka 1988). On the subject of outcasts in the Japanese mafia, Herbert (2000, 149) writes, "Although no reliable data exist, it is well known that these two excluded minorities are greatly over-represented in *yakuza* syndicates.... According to an unofficial police estimate, around 10 percent of Yamaguchi-gumi members are of Korean origin, while up to 70 percent of them may have a Burakumin [sic] background."

22. As a spry, elderly gentleman put it one day when I approached him in the Forest Park and listened to him describe the "smoke" *(kemuri)* that he said came from the facility: "All these people around you here. They don't believe in Azuma Disease. And all the other people [who believe in Azuma Disease], they're long gone!"

23. Several official meetings, such as one ward hearing in October 1999, were punctuated by commission members berating officials of the Tokyo waste bureau for egregious manipulations of statistical data related to toxic levels in Izawa and other transparent failed attempts to undermine GEAD's assertions.

Chapter 7: Growth, Sex, Fertility, and Decline

1. The oil shocks of 1973–1974, and the brief recession to which they gave rise, were viewed as an aberration by most Japanese (Lincoln 1988, 26–34).

2. Indeed, after visiting factories of the day, one legislator wondered whether Japan's population would decline given such conditions and such exploitation of child labor (Hane 1998, 160).

3. As Thomas Havens observes, "For many years, Japan had told the world she needed more territory for her crowded population" (1978, 135), and yet once given the opportunity, Japan zealously promoted population expansion in the new territories.

4. Of course, there remains intense controversy regarding whether Emperor Hirohito served as figurehead or as active helmsman—a "dynamic emperor" who was "a real war leader" (Bix 2000, 12)—of the militarist state.

5. It is worth recalling, however, that invocations of Japan's "hundred million" included millions of the empire's foreigners, outcasts, and so on, who themselves could be viewed as contaminating elements (e.g., Ohnuki-Tierney 2002).

6. The fertility rate in Japan, calculated as the average number of children per woman of childbearing age, slid to 1.38 in 1998 (Foreign Press Center 2000), the year I embarked upon field research. At least 2.08 is needed to achieve population stability. By 2004, this figure had fallen to 1.29. The population of Japan is predicted to plummet from 127 million to just over 100 million by 2050. Meanwhile, Japan has a very high percentage of elderly—Japanese over 65 comprise nearly 20 percent of the population (Foreign Press Center 2006, 13–18). Economists and social commentators have warned against a shrinking workforce, which will eventually lead either to a sustained contraction of the world's second-largest economy (and partial reversal of the Japanese "economic miracle") or to a massive influx of foreign workers, who will in turn threaten Japan's much-fabled social and linguistic (and racial) "homogeneity."

7. In a sense, some *frītā*, or NEETs (a common Japanese acronym for Not in Education, Employment, or Training), engage in what may be called self-garbaging—junking to some extent their prospects in mainstream society by exiting the standard tracks and participating in new networks of alternative relation and status. Not surprisingly, this social choice is frequently viewed as a "waste" by those committed to more establishment value systems, such as parents.

8. As noted earlier, *shimaguni* means "island country"; it is used to explain Japan's isolation from the world and, by extension, the basis of Japan's Japa-

neseness. *Konjō* means "disposition." The term *shimaguni konjō* describes the petty rivalries and sniping in close quarters that comprise the opposite of normative constructions of *furusato* community cheer. One source of local difficulty for Mama-san came from the proprietress (and some patrons) of a far less prosperous bar just up the road. Considerably more established in the community, they vocally resented Mama-san's arrival and her conspicuous success subsequently.

9. Some Sugibayashi-based informants voiced concern that not only did they live with an incinerator in the center of their own community but that across the nearby ward border, another equally large incinerator was also in operation. These informants noted correctly that while one incinerator might meet environmental restrictions, there was no extensive official consideration of the combined emissions of the two.

10. According to a nurse at the local OB-GYN hospital, defects did occur, but in a society where families are known to sequester retarded or acutely disabled children in special homes to avoid shame by association, such defects would not likely surface easily in gossip or most ethnographic interviews.

11. It is important to remember, of course, that miscarriages are more numerous in every society than commonly recognized and often remain shrouded in secrecy. I have no reason to believe that the number of miscarriages in Horiuchi was higher than elsewhere in urban Japan.

12. My question had not specifically referenced dioxins, but for quite some time we had been discussing dioxins and other toxic pollutants in the community.

13. Another Chikurin regular, Moritake-san, backhandedly proclaimed Katsu-san a "genius at *shimaguni konjō*" (*shimaguni konjō no tensai*), or petty backstabbing, one evening when I asked some locals at the bar to explain the term, and he nominated Katsu-san to define it. Katsu-san got the message and declined his invitation.

14. Tōda-san said nothing, endeavoring to stay above the fray; he explained to me later, "I don't want to get involved in that. . . ."

15. PCBs are endocrine disruptors similar to dioxin. Astonishingly, Taka-chan articulates the full name in scientific English.

16. Shin-chan was alluding to the dreaded, much publicized problem of "sick house" (*shikku hausu*), as well as to toxic pollution outdoors.

17. The EPA report (2000) provides a long list of scientific studies on both animals and humans, but the implications for human health specifically are very convincing.

18. See Michael Ashkenazi's account (Ashkenazi and Rotenberg 1999) of his experiences in all-male sections of Japanese bathhouses. These will certainly sound familiar to male foreigners who have immersed themselves in Japan's bath culture.

19. Implicated in this suspicion of foreign sexual mystique, real or imag-

ined, was a critique of young women's increased independence, including sexual independence. As evidenced by the broad-based outrage provoked by late-1980s media reports about young female Japanese going on trips abroad to find sexual adventure with foreign men—dubbed "yellow cabs" for how apparently easy they were to pick up (Ieda 1991)—older people in Japanese society had decidedly mixed feelings about these young women and their challenge to the established order.

20. Some of the confusion between impeded testicular and penis development resulted from the vagueness of the term chosen for discussion, *seishoku-ki,* which indexes "reproductive organs" rather than specifying one or the other.

21. The study determined that the average weight of testes increased from around 17 grams in 1964 to slightly more than 19 grams in 1980 before decreasing to about 18 grams since the early 1990s. It was unclear from the findings whether the JEA took into account the well-known increase in overall body size of Japanese males over the postwar period as a result of enhanced diet. Organs with higher fat content showed high levels of dioxin—unsurprising since dioxin is often stored in fat cells. The brain, on the other hand, showed extremely low levels (Japan Environment Agency 1999b).

22. This repercussion of exposure has been closely documented in Seveso, Italy, among other notorious areas with long-term contamination (e.g., Mocarelli, Gerthoux, et al. 2000). Analyses of Japanese regions have discovered similar, though less dramatic, findings (Matsuzaki and Yamazaki 2000; Kamiyama, Matsuzaki, et al. 2000).

23. Prince Naruhito's brother Akishino, the next in line for the throne, and his wife later gave birth to a boy in 2006.

24. These measures echoed ones initiated by the pronatalist imperial government in the 1930s. Daycare facilities for the farming season were increased nearly fourfold between 1933 and 1938 (Miyake 1991). And in the 1940s, the state offered free education to families with over ten children (Havens 1978).

25. To be sure, the cabinets, ministries, agencies, and other institutions that comprise a government—not to mention the separate administrative and regulatory edifices of major cities such as Tokyo—are rife with factions and political conflict over competing agendas. Rest assured that I do not view the Japanese government as a monolithic state entity handing down dictates from on high that are generated by a harmonious, coherent command structure.

26. http://www.komei.or.jp/en/policy/major_environmental.html for the English version (accessed September 2009).

27. A bureaucrat in Tokyo's waste management section later confirmed, for his part, that national and metropolitan efforts to limit dioxins sharpened in 1997, partly due to reproductive and public health concerns, and became more urgent with the media frenzy surrounding dioxins in early 1999.

28. It is common to put harmful dioxin-like substances like furans within

the purview of official toxic studies as well, as the UNEP does, but I simplify by using the term "dioxins" here.

29. In fact, some informants at Chikurin called the gingko nut, a traditional snack served there, "Japanese Viagra." Gingko improves circulation, and they swore it had the same effect as Viagra.

30. While Allison's ethnography focuses on the rarefied plane of hostess clubs, so expensive that they are usually patronized only by employees of large institutions on the company dime, there exist numerous lower-strata establishments that are appropriate for employees of businesses with different budgets (e.g., Mackie 1988).

Chapter 8: Constructing "Sustainable" Japan

Epigraph: Song cited in Huddle and Reich 1987, 224–225.

1. Indeed, the part of Wakasu now covered with the golf course was earlier called the New Island of Dreams *(shin-yume no shima)* (Endō 2004, 799).

2. http://wakasu.golftk.com/history.html (accessed April 2010). Naturally, the marine sustainability of reclaiming large tracts of land in Tokyo Bay, and filling the space within their concrete barriers with a mountain of waste and dirt, is not normally trumpeted by the institutions that manage these artificial islands. (Indeed, toward the center of the enormous bay is the spot where, on 2 September 1945, the Japanese surrendered to the Allies on the deck of the USS *Missouri*. Since that date, reclaimed land has covered approximately half the distance to that point, raising the theoretical possibility that Japan will be able to "domesticate" the event with garbage-packed land there in future decades. The capital's wetlands and coastline have been subject to aggressive land reclamation since the city's foundation (see Endō 2004).

3. One only has to consider that it no longer takes an ax to cut down a tree, even in the developing world—available tools run from simple chainsaws to free-standing machines that can topple a large tree in seconds. Third World populations also turn to swidden burning of forest for its ease, simplicity, and brutal economy.

4. I use the past tense here because in 2001 these bureaucratic entities were reshuffled and in some cases renamed. I return to these changes in a coming section.

5. Japan merits some sympathy in this regard, since its leaders and emissaries have certainly expressed deep regret to their former enemies and imperial subjects many times over past decades. But other actions—the spurning of claims by "comfort women" and more or less enslaved foreign war laborers, the repeated tone-deaf revisions of school textbooks that bowdlerize Japanese atrocities, the pandering visits to honor war dead at Japan's politically sensitive Yasukuni Shrine, and numerous other attempts to gloss over the horrors committed by Japan's military regime—create a strong impression of stub-

bornness, even rejection of responsibility for Japan's reviled wartime conduct. Cynical political manipulation of these issues by nations such as China, however, makes it clear that there is plenty of blame to spread around in this regard.

6. The convention was agreed to in 1982 at the IWC's 34th Annual Meeting, with the moratorium eventually enshrined in paragraphs 10 and 10(e) of the Schedule of the Convention and coming into force during 1985–1986.

7. See http://www.iwcoffice.org/conservation/table_permit.htm (accessed April 2010).

8. The seafood companies that used to participate in whale meat distribution sold all their stock in their shared enterprise to institutions loosely linked to the government in 2006, demonstrating how "unsustainable" demand for whale meat was. Referring to international antiwhaling opinion, one company board member stated, "As a company selling seafood internationally, being involved in whale [meat] is completely not a good thing *(kujira ni kakawatte yoi koto ha mattaku nai)*" (*Asahi Shimbun,* 13 June 2008). The company's managing director added, "For people who ate it a long time ago, whale probably seems nostalgic, but other meat tastes better," and the managing director of another company summed up the problem: "Young people don't eat [it]" (*Asahi Shimbun,* 13 June 2008; see also *Asahi Shimbun,* 2 February 2008).

9. E.g., http://www.iwcoffice.org/conservation/estimate.htm (accessed April 2010).

10. See http://www.icrwhale.org/JARPApaper.htm (accessed April 2010) for a long list of such studies, many unpublished. (Cf. International Whaling Commission Scientific Committee 2006.)

11. Though the Spartan mystique of the warrior-poet has been romanticized in popular literature on Japan (e.g., Clavell 1975), the fact that Tokugawa felt the need to create rules against extravagance probably indicates that his vassals may not always have resisted self-indulgence on their own.

12. Edward Seidensticker writes that in Edo, farmers were willing to pay higher rates for the excreta of more monied households: "The upper-class product was richer in nutriment, apparently. So, apparently, was male excrement" (Seidensticker 1991a, 283). John Embree corroborates Japan's excremental reverence in the 1930s in his rural village fieldsite, stating that "every scrap of human manure is used" and that "[t]he school and village office rent out the right to collect their night-soil" (Embree 1946 [1939], 35, 43; cf. Knight 1997).

13. During the approximately 270 years of the Edo period, the city reclaimed 2,700 hectares of land, and from the Meiji period to the present (approximately 140 years), Tokyo reclaimed 6,000 hectares (Endō 2004).

14. Such diseases plagued most urban concentrations of the day, in Asia as in Europe (Macfarlane 1997 cites a range of relevant historical sources).

15. However, by the late-Meiji period, in this growing city, "the unremitting

smell of burning garbage is a detail commonly remarked upon" (Seidensticker 1991a, 282), a foreshadowing of contemporary massive-scale incineration in the capital.

16. Some of this use grew out of desperation. For example, Japan used whale skin toward the end of World War II as a substitute for leather (Havens 1978, 16).

17. Shin-chan finally burst out, "Which people in the world kill the most whales? The Americans!... The Inuit [in Alaska]!" (In this, Shin-chan was greatly misinformed.) Japanese informants, by and large, seemed to be well armed with such cross-cultural arguments, whether accurate or otherwise, given the criticism that Japan endured from much of the rest of the world and the appearance of stories on whaling in mass media.

18. Later in the postwar period, though, frugality and reserve generally came to be viewed with nostalgia, with older residents criticizing some of the changes that their communities had undergone (R. Smith 1978, 114, 177–182; Dore 1978, 90–91).

19. Another moral dimension of waste was far less discussed. Japanese train stations and other areas sporadically offered locked white disposal containers (*shiroi posuto*, literally "white post box") for pornographic magazines, books, video cassettes, and DVDs. Communities urged adult users to dump porn *(poruno)* to keep it out of the hands of minors, and the white boxes allowed closet readers to junk their train reading, for example, before returning home. But the adult material could also be left carelessly elsewhere.

20. Mujirushi, which became its own company in 1989, has expanded considerably and now can be found in cities all over Japan, including approximately fifty-seven locations in Tokyo alone.

21. It would be inaccurate in this case to use the term "landlord," since Mr. Iwasaki made it clear that such menial household matters as signing the lease, collecting rent, and so on were the province of the landlady—traditionally a gendered role in Japan.

22. Perhaps anticipating these responses, Tokyo authorities tried—ironically—to use the new concern over dioxins to enforce better waste separation and recycling habits. They distributed pamphlets clarifying the manner in which dioxins are released from waste incineration and the toxins' deleterious effects; the pamphlets urge greater vigilance in separating plastics from burnable waste collections and encourage recycling (Tokyo Bureau of Environment 1999; Japan Environment Agency 1999d; Japan Environment Agency 1999e).

23. While the huge room was filled with a horseshoe-shaped table for the committee members and row upon row of assigned tables and seats for Tokyo waste bureau members, Tokyo metropolitan officials, ward assembly members, journalists, and so on (with many empty desks and chairs along the way), the representatives of the GEAD and their friends were forced to perch in a crowded single row of temporary chairs at the very back of the long room. Their

access was also closely monitored, with all visitors forced to register downstairs and wear special badges.

24. Despite this belated acknowledgement of toxic exposure in the community, Izawa protesters and their supporters bitterly opposed this official version. First of all, it allowed the facility to continue operations. In addition, protesters claimed, the official conclusion ignored the numerous dangerous toxins repeatedly measured in the air around the facility.

25. Cynical observers (such as myself) might indeed speculate as to whether this overcapacity was intentional.

26. VOCs include many dangerous airborne pollutants from industrial and other chemicals that include the indoor toxins involved in "sick house" syndrome as well as those found outdoors. Fuels, solvents, refrigerants, coatings, and so on, are frequently VOCs. These tend toward a gaseous state and are therefore difficult to control.

27. Taking into account the Tokyo Metropolitan Government's manipulation of official toxic waste statistics, which I described in Chapter 3, and the pressures on the Japanese state with regard to toxins, waste, and environmental problems generally, perhaps it would be safest to take any such figures with a grain of salt.

Conclusions

1. See Fortun 2001, 52; Douglas and Wildavsky 1982; Cronon 1996.

2. These environmentalists were trying to limit demand for Siberian lumber by Japanese construction companies, for example.

3. Take, for example, the sustained impact of Native American interventions on the ecology of Yosemite in California, including the large-scale forest fires they set (see Schama 1995).

References

Aldrich, Daniel P. 2005. Leviathan or agile state? Strategies and tool kits for siting public bads. Ph.D. diss., Political Economy and Government, Kennedy School of Government, Harvard University.

———. 2008. *Site fights: Divisive facilities and civil society in Japan and the West.* Ithaca, NY: Cornell University Press.

Allison, Anne. 1994. *Nightwork: Sexuality, pleasure, and corporate masculinity in a Tokyo hostess club.* Chicago: University of Chicago Press.

Amano Reiko. 2001. *Damu to Nihon* (Dams and Japan). Tokyo: Iwanami Shinsho.

ANRE. 2008. *Energy in Japan 2008.* Tokyo: Agency for Natural Resources and Energy.

———. 2009. *Annual energy report 2008 (Outline).* Tokyo: Agency for Natural Resources and Energy.

Appadurai, Arjun, ed. 1988. *The social life of things: Commodities in cultural perspective.* Cambridge: Cambridge University Press.

Architectural Institute of Japan. 2002. *Shikku hausu taisaku no baiburu.* (The bible of sick house countermeasures). Tokyo: Shōkokusha.

Asahi Shimbun. 1999, 10 February. Deeta kōgyō anzen uttae: JA-Tokorozawa no nosai daiokishin chōsa (Pleading safe in public presentation of data: Tokorozawa agricultural cooperative's vegetable dioxin survey). Evening edition, p. 31.

———. 1999, 22 February. Ro dake yasunde gomi no yama: "Sanpai Ginza" Tokorozawa (With only the incinerators taking a break, there's a mountain of garbage: Tokorozawa the "Industrial Waste Ginza"). Morning edition, p. 3.

———. 1999, 26 March. "Shōkyakujō de daiokishin higai" moto jyūgyōin ga rōsai shinsei (Former incinerator employees who were "victims of dioxin poisoning" apply for workers compensation). Evening edition, p. 1.

———. 1999, 27 March. Ketsuchū ni kōnōdo daiokishin: Osaka, Nose no shōkyakujyō rōdōsha 92-nin chōsa (High dioxin levels in blood: Survey of 92 employees at Osaka's Nose incinerator). Morning edition, p. 1.

———. 1999, 6 August. Karasu: Dō suru anata nara...? Kashikosa yue ni kirawareru. (*Karasu:* What would you do...? They're clever, therefore they're disliked). Morning edition, p. 29.

———. 1999, 7 August. Karasu: Dō suru anata nara...? Kyōson no michi doko ni aru? (*Karasu:* What would you do...? Where is the road of coexistence?). Morning edition, p. 25.

———. 2008, 2 February. Chōsa hogei, futokoro mo pinchi; kuni kara no yūshi jūokuen kaesezu (Research whaling also feeling it in the purse; Unable to repay one billion yen loan from the government).

———. 2008, 13 June. Shōgyōhogei saisannyū, suisan ōte sansha ha hitei: "Yoi koto nai" (Three big fisheries companies reject continuation of whaling business, saying: "It's not a good thing"). Written by Oyamada Kenji.

Ashkenazi, Michael, and Robert Rotenberg. 1999. "Cleansing cultures: Public bathing and the naked anthropologist in Japan and Austria." In *Sex, sexuality, and the anthropologist,* eds. Fran Markowitz and M. Ashkenazi. Champaign, IL: University of Illinois Press.

Askew, David. 2002. The debate on the "Japanese" race in imperial Japan: Displacement or coexistence? *Japanese Review of Cultural Anthropology* 3:79–96.

———. 2004. Debating the "Japanese" race in Meiji Japan: Towards a history of early Japanese anthropology. In *The making of anthropology in East and Southeast Asia,* eds. S. Yamashita, J. Bosco, and J. S. Eades. Oxford: Berghahn.

Asquith, Pamela, and Arne Kalland, eds. 1997. *Japanese images of nature: Cultural perspectives.* Richmond, Surrey: Curzon Press.

Associated Press. 2007, 15 March. Not tonight, dear.

Baba K., T. Iwamoto, et al. 2000. Nihonjin jyakunen-dansei ni okeru seieki-shoken to sono kisetsu hendō (The reproductive function of Japanese young men: Semen quality and its seasonal variation). Paper presented at the Third Annual Meeting of the Japan Society of Endocrine Disruptor Research, 15 December 2000, Yokohama, Japan.

Bank of Japan. 1998. *Kigyō tanki keizai kansoku chōsa* (Tankan: Short-term economic survey of enterprises in Japan). 4th quarter report. Bank of Japan.

———. 1999. *Kigyō tanki keizai kansoku chōsa* (Tankan: Short-term economic survey of enterprises in Japan) 4th quarter report. Bank of Japan.

Baudrillard, Jean. 1998 (1970). *The consumer society: Myths and structures.* London: Sage.

Beck, Ulrich. 1992. *Risk society: Towards a new modernity.* Trans. Mark Ritter. London: Sage Books.

Befu Harumi, ed. 1993. *Cultural nationalism in East Asia: Representation and identity.* Berkeley: Institute of East Asian Studies, University of California.

―――. 1997. Watsuji Tetsurō's ecological approach: Its philosophical foundation. In *Japanese images of nature: Cultural perspectives,* eds. P. Asquith and A. Kalland. Richmond, Surrey: Curzon Press.

―――. 2001. *Hegemony of homogeneity: An anthropological analysis of nihonjinron.* Melbourne: Trans Pacific Press.

Benedict, Ruth. 1947. *The chrysanthemum and the sword: Patterns of Japanese culture.* London: Secker & Warburg.

Berglund, Eeva. 1998. *Knowing nature, knowing science: An ethnography of local environmental activism.* Cambridge: White Horse Press.

Berque, Augustin. 1997a (1986). *Japan: Nature, artifice and Japanese culture.* Trans. R. Schwarz. Yelvertoft Manor, Northamptonshire: Pilkington Press.

―――. 1997b (1993). *Japan: Cities and social bonds.* Trans. C. Turner. Yelvertoft Manor, Northamptonshire: Pilkington Press.

Bertazzi, Pier, Ilaria Bernucci, et al. 1998. The Seveso studies on early and long-term effects of dioxin exposure: A review. *Environmental Health Perspectives* 106 (Suppl 2):625–633.

Bestor, Theodore. 1989. *Neighborhood Tokyo.* Stanford: Stanford University Press.

―――. 2001. Supply-side sushi: Commodity, market, and the global city. *American Anthropologist* 103 (1):76–95.

Bird, S. Elizabeth. 1992. *For enquiring minds: A cultural study of supermarket tabloids.* Knoxville: University of Tennessee Press.

―――. 2003. *The audience in everyday life: Living in a media world.* New York: Routledge.

Bix, Herbert P. 2000. *Hirohito and the making of modern Japan.* New York: HarperCollins.

Bondy, Christopher. 2005. Becoming burakumin: Education, identity and social awareness in two Japanese communities. Ph.D. diss., Department of Sociology, University of Hawai'i at Manoa.

Bourdieu, Pierre. 1998. *On television and journalism.* Trans. Priscilla Parkhurst Ferguson. London: Pluto Press.

Bowring, Richard, and Peter Kornicki, eds. 1993. *The Cambridge encyclopedia of Japan.* Cambridge: Cambridge University Press.

Broadbent, Jeffrey. 1998. *Environmental politics in Japan: Networks of power and protest.* Cambridge: Cambridge University Press.

Brockington, Daniel. 2002. *Fortress conservation: The preservation of the Mkomazi Game Reserve, Tanzania.* Oxford: James Currey.

Brown, Mary Ellen. 1989. Soap opera and women's culture: Politics and the popular. In *Doing research on women's communication: Perspectives on theory and method,* eds. Kathryn Carter and Carole Spitzack. Norwood, NJ: Ablex.

Buckley, Thomas, and Alma Gottlieb, eds. 1988. *Blood magic: The anthropology of menstruation.* Berkeley: University of California Press.

Carrier, James G. 2004. Introduction. In *Confronting environments: Local understanding in a globalizing world,* ed. J. G. Carrier. Walnut Creek, CA: AltaMira.

Casper, Monica, ed. 2003. *Synthetic planet: Chemical politics and the hazards of modern life.* New York: Routledge.

Castells, Manuel. 2004. *The power of identity.* Oxford: Blackwell.

Chapin, Mac. 2004. A challenge to conservationists. *World Watch* 17 (6):17–31.

Clammer, John. 1997. *Contemporary urban Japan: A sociology of consumption.* Oxford: Blackwell.

Clapham, Phillip J., Simon Childerhouse, et al. 2006. The whaling issue: Conservation, confusion, casuistry. *Marine Policy.*

Classen, Constance. 1993. *Worlds of sense: Exploring the senses in history and across cultures.* London: Routledge.

Clavell, James. 1975. *Shogun.* London: Hodder & Stoughton.

Colborn, Theo, Dianne Dumanoski, and John Peterson Myers. 1996. *Our stolen future: Are we threatening our fertility, intelligence, survival? A scientific detective story.* Boston: Little, Brown.

Conner, Walker. 1978. A nation is a nation, is a state, is an ethnic group, is a... *Ethnic and Racial Studies* 1 (4):379–388.

Crawcour, E. Sydney. 1988. Industrialization and technological change, 1885–1920. In *The Cambridge history of Japan.* Vol. 6: *The twentieth century,* ed. Peter Duus. Cambridge: Cambridge University Press.

Cronon, William, ed. 1996. *Uncommon ground: Rethinking the human place in nature.* New York: Norton.

Cybriwsky, Roman. 1998. *Tokyo: The shogun's city at the 21st century.* Chichester, UK: Wiley.

Dale, Peter. 1986. *The myth of Japanese uniqueness.* New York: St. Martin's Press.

Dauvergne, Peter. 1997. *Shadows in the forest: Japan and the politics of timber in Southeast Asia.* Cambridge, MA: MIT Press.

Davis, John H., Jr. 2000. Blurring the boundaries of the buraku(min). In *Globalization and social change in contemporary Japan,* eds. J. Eades, T. Gill, and S. Yamashita. Melbourne: Trans Pacific Press.

Delaney, C., and S. Yanagisako, eds. 1995. *Naturalizing power: Essays in feminist cultural analysis.* New York: Routledge.

Delegation of Japan. 2008. Briefing note. 60th annual meeting of the International Whaling Commission, 23–27 June 2008, in Santiago, Chile. Tokyo: Institute for Cetacean Research. (http://www.icrwhale.org/60MediaKit.htm)

Deliége, Robert. 1999. *The untouchables of India.* Trans. Nora Scott. Oxford: Berg.

DeVos, George, and Hiroshi Wagatsuma, eds. 1966. *Japan's invisible race: Caste in culture and personality.* Berkeley: University of California Press.

Diamond, Jared. 2005. *Guns, germs and steel: A short history of everybody for the last 13,000 years.* Rev. ed. New York: Norton.

Dinmore, Eric. 2006. A small island nation poor in resources: Natural and human resource anxieties in trans-World War II Japan. Ph.D. diss., Princeton University.

Dore, Ronald. 1973 (1958). *City life in Japan: A study of a Tokyo ward.* Berkeley: University of California Press.

——. 1978. *Shinohata: A portrait of a Japanese village.* London: Allen Lane.

Douglas, Mary. 1966. *Purity and danger: An analysis of the concepts of pollution and taboo.* London: Routledge & Kegan Paul.

——, and Aaron Wildavsky. 1982. *Risk and culture: An essay in the selection of technical and environmental dangers.* Berkeley: University of California Press.

Dower, John W. 1986. *War without mercy: Race and power in the Pacific War.* New York: Pantheon.

——. 1999. *Embracing defeat: Japan in the wake of World War II.* New York: Norton.

DPJ. 2009. *DPJ manifesto 2009.* Tokyo: Democratic Party of Japan.

Dumont, Louis. 1980 (1967). *Homo hierarchicus: The caste system and its significance.* Trans. Mark Sainsbury, Louis Dumont, and Basia Gulati. Rev. ed. Chicago: University of Chicago Press.

Durkheim, Emile. 1995 (1912). *The elementary forms of the religious life.* Trans. Karen E. Fields. New York: The Free Press.

Duus, Peter. 1976. *Feudalism in Japan.* Rev. ed. New York: Knopf.

Eades, Jerry, Tom Gill, and Shinji Yamashita, eds. 2000. *Globalization and social change in contemporary Japan.* Melbourne: Trans Pacific Press.

Edelstein, M. 1988. *Contaminated communities: The social and psychological impacts of residential toxic exposure.* Boulder, CO: Westview Press.

Embree. 1946 (1939). *Suye Mura: A Japanese village.* London: Kegan Paul.

Endō Takeshi. 2004. Tōkyō rinkaibu ni okeru umetate no rekishi (Historical review of reclamation works in the Tokyo Port Area). *Journal of Geography* 113 (6):785–801.

EPA. 2000. Exposure and human health reassessment of 2,3,7,8-Tetrachlorodibenzo-p-Dioxin (TCDD) and related compounds. Draft final report. Washington, DC: US Environmental Protection Agency.

Escobar, Arturo. 1998. Whose knowledge? Whose nature? Biodiversity, conservation, and the political ecology of social movements. *Journal of Political Ecology* 5:53–82.

——. 1999. After nature: Steps to an antiessentialist political ecology. *Current Anthropology* 40:1–30.

Ezawa, Aya. 2006. How Japanese single mothers work. *Japanstudien* 18:57–82.

Fairhead, James, and Melissa Leach. 1996. *Misreading the African landscape.* Cambridge: Cambridge University Press.

Farmer, Paul. 1992. *AIDS and accusation: Haiti and the geography of blame.* Berkeley: University of California Press.

Foreign Press Center. 2000. *Facts and figures of Japan.* Tokyo: Foreign Press Center Japan.

———. 2006. *Facts and figures of Japan.* Tokyo: Foreign Press Center Japan.

Fortun, Kim. 2001. *Advocacy after Bhopal: Environmentalism, disaster, new global orders.* Chicago: University of Chicago Press.

Fowler, Edward. 1996. *San'ya blues: Laboring life in contemporary Tokyo.* Ithaca, NY: Cornell University Press.

Garon, Sheldon. 1997. *Molding Japanese minds: The state in everyday life.* Princeton: Princeton University Press.

Geertz, Clifford. 1963. The integrative revolution: Primordial sentiments and civil politics in the new states. In *Old societies and new states: The quest for modernity in Asia and Africa,* ed. C. Geertz. New York: The Free Press.

Gellner, Ernest. 1983. *Nations and nationalism.* Oxford: Blackwell.

Genda Yuji. 2004. *Jobu kurieishon* (Job creation in Japan) Tokyo: Nihon Keizai Shimbunsha.

George, Timothy S. 2001. *Minamata: Pollution and the struggle for democracy in postwar Japan.* Cambridge, MA: Harvard University Press.

Gill, Tom. 2000. *Yoseba* and *ninpudashi:* Changing patterns of employment on the fringes of the Japanese economy. In *Globalization and social change in contemporary Japan,* eds. J. Eades, T. Gill, and S. Yamashita. Melbourne: Trans Pacific Press.

———. 2001. *Men of uncertainty: The social organization of day laborers in contemporary Japan.* Albany: State University of New York Press.

———. 2005. The *kegare* category: Ritual pollution and social discrimination in contemporary Japan. Paper presented at the Anthropologists of Japan in Japan (AJJ) conference, 5–6 November 2005, at Sophia University, Tokyo.

Government of Japan. 2006. Japan's fourth national communication under the United Nations Framework Convention on Climate Change. Tokyo: Cabinet Office.

Hane, Mikiso. 1998. The textile factory workers. In *Meiji Japan: Political, economic and social history, 1868–1912.* Vol. 2: *The growth of the Meiji state,* ed. P. Kornicki. London: Routledge.

Hardacre, Helen. 1997. *Marketing the menacing fetus in Japan.* Berkeley: University of California Press.

Harris, Ruth L. 1989. The meanings of "waste" in Old and Middle English. Ph.D. diss., University of Washington.

Hasegawa Miki. 2006. *We are not garbage! The homeless movement in Tokyo, 1994–2002.* New York: Routledge.

Hauser, William B. 1991. Women and war: The Japanese film image. In *Recreating Japanese women, 1600–1945,* ed. G. Bernstein. Berkeley: University of California Press.

Havens, Thomas R. H. 1978. *Valley of darkness: The Japanese people and World War Two.* New York: Norton.

Hawkins, Gay. 2004. Shit in public. *Australian Humanities Review* 31–32.

Heidegger, Martin. 1988 (1927). *Being and time.* Trans. John Macquarrie and Edward Robinson. Oxford: Blackwell.

Hendry, Joy. 1993. *Wrapping culture: Politeness. presentation, and power in Japan and other societies.* Oxford: Clarendon.

———. 1997. Nature tamed: Gardens as a microcosm of Japan's view of the world. In *Japanese images of nature: Cultural perspectives,* eds. P. Asquith and A. Kalland. Richmond, Surrey: Curzon Press.

Herbert, Wolfgang. 2000. The *yakuza* and the law. In *Globalization and social change in contemporary Japan,* eds. J. Eades, T. Gill, and S. Yamashita. Melbourne: Trans Pacific Press.

Hetherington, Kevin. 2004. Second-handedness: Consumption, Disposal and Absent Presence. *Environment and Planning D: Society and Space* 22 (1):157–173.

Hoffman, Yoel. 1986. *Japanese death poems.* Rutland, VT: Tuttle.

Horie Kunio. 1984. *Genpatsu jipushii* (Nuclear gypsy). Tokyo: Kōdansha

Hoshino Takanori. 2008. Transition to municipal management: Cleaning human waste in Tokyo in the modern era. *Japan Review* 20:189–202.

Howes, David, ed. 1991. *The varieties of sensory experience: A sourcebook in the anthropology of the senses.* Toronto: University of Toronto Press.

Huddle, N., and M. Reich. 1987. *Island of dreams: Environmental crisis in Japan.* Cambridge, MA: Schenkman Books.

Ieda Shōko. 1991. *Ierō kyabu: Narita wo tobitatta onnatachi* (Yellow cabs: The girls who took off from Narita). Tokyo: Kōyūsha.

Igoe, Jim. 2003. Scaling up civil society: Donor money, NGOs and the pastoralist land rights movement in Tanzania. *Development and Change* 34:863–885.

———. 2005. Global indigenism and spaceship earth: Convergence, space, and re-entry friction. *Globalizations* 2:377–390.

Imai, Masaaki. 1986. *Kaizen: The key to Japan's competitive success.* New York: McGraw-Hill/Irwin.

Imamura, A. E. 1987. *Urban Japanese housewives at home and in the community.* Honolulu: University of Hawai'i Press.

Independent, The. 2001, 17 April. Pregnant princess raises Japanese hopes for an heir. By Richard Lloyd Parry.

Ingold, Tim. 1993. Globes and spheres: The topology of environmentalism. In *Environmentalism: The view from anthropology,* ed. K. Milton. London: Routledge.

Inoue Masao. 2004. *Shikku hausu no bōshi to taisaku* (Sick house prevention and counter-measures). Tokyo: Nikkan Kōgyō Shimbunsha.

Institute for Global Environmental Studies. 1999. Quality of the environment in Japan 1999: Environmental messages toward sustainable development in the 21st century. Kanagawa: Institute for Global Environmental Studies.

Institute of Medicine of the National Academies. 2003. *Dioxins and dioxin-like compounds in the food supply: Strategies to decrease exposure.* Washington, DC: National Academies Press.

International Whaling Commission. 1986a. Chairman's report of the thirty-seventh annual meeting. *Reports of the International Whaling Commission* 36:20.

———. 1986b. Report of the scientific committee. *Reports of the International Whaling Commission* 36:31–32.

———. 1989. Report of the special meeting of the Scientific Committee to consider the Japanese research permit (feasibility study). *Reports of the International Whaling Commission.* 39:159–164.

International Whaling Commission Scientific Committee. 2006. Report of the intersessional workshop to review data and results from special permit research on minke whales in the Antarctic, Tokyo 4–8 December 2006. (http://www.iwcoffice.org/_documents/conservation/SC-59-Rep1.pdf)

Iriye Akira. 1989. Japan's drive to great-power status. In *The Cambridge history of Japan.* Vol. 5: *The nineteenth century,* ed. Marius Jansen. Cambridge: Cambridge University Press.

Ishikawa S. et al. (eds). 2003. *Watashi no kagaku busshitsu kabinshō: Kanshata-chi no kiroku* (My chemical hypersensitivity: Patients' journals). Published in Japanese. Publication of the Association of Patients of Chemical Hypersensitivity. Tokyo: Jissensha.

Ivy, Marilyn. 1995. *Discourses of the vanishing: Modernity, phantasm, Japan.* Chicago: University of Chicago Press.

Japan Environment Agency. 1994. *The basic environment plan.* Tokyo: Japan Environment Agency.

———. 1997a. *The national action plan for Agenda 21.* Tokyo: Japan Environment Agency.

———. 1997b. Statement from Prime Minister of Japan Ryutaro Hashimoto at UNGASS. *Japan Environment Quarterly* 2 (3):3–8.

———. 1999a. *Our intensive efforts to overcome the tragic history of Minamata Disease.* Tokyo: Japan Environment Agency.

———. 1999b. *Daiokishin-rui no jintai ketsueki yasei dōbutsu oyobi shokuji-chū no chikuseki jōkyō to ni tsuite* (On the state of accumulation of dioxins in human blood, wildlife, and in food). Tokyo: Japan Environment Agency.

———. 1999c. *Basic guidelines of Japan for the promotion of measures against dioxins (revised draft).* Report of the Ministerial Council on Dioxin Policy Japan. Tokyo: Japan Environment Agency.

———. 1999d. *Daiokishin-rui wo herazu* (Not reducing dioxins). Tokyo: Japan Environment Agency.

———. 1999e. *Daiokishin-tte nāni?* (What are dioxins?). Tokyo: Japan Environment Agency.

Japan Institute of Energy. 2002. *Baiomasu handobukku* (Biomass handbook). Tokyo: Japan Institute of Energy.

Japan Times. 2000, 2 July. City man vs. nature: Garbage, indifference fuelling crow plague. By Taiga Uranaka.

———. 2000, 12 November. Japan's not-so-silent media conspiracy. By Roger Pulvers.

———. 2007, 3 November. Plastic incineration draws ire: Environmentalists unswayed by limited tests, fear risks. By E. Prideaux.

———. 2008, 14 August. Shock tactics return: Art collective Chim↑Pom provokes Tokyo with a crow and a rat. By Edan Corkill.

Jay, Martin. 1993. *Downcast eyes: The denigration of vision in twentieth-century French thought.* Berkeley: University of California Press.

JCO Criticality Accident Comprehensive Assessment Committee. 2002. *JCO rinkai jiko: Sannen-go ni miete kita mono* (The JCO criticality accident: Three years after). Tokyo: Citizens' Nuclear Information Center.

Jeremy, M., and M. E. Robinson. 1989. *Ceremony and symbolism in the Japanese Home.* Manchester: Manchester University Press.

Jinnai Hidenobu. 1995 (1985). *Tokyo: A spatial anthropology.* Trans. Kimiko Nishimura. Berkeley: University of California Press.

——— et al. 1989. *Edo-Tokyo no mikata shirabekata.* (*Edo-Tokyo: Searching and seeing.* Published in Japanese.) Tokyo: Kashima Shuppankai.

Joy, Bill. 2000. Why the future doesn't need us. *Wired* 8 (4):238–262.

Kalland, Arne. 1982. *Shingu: A study of a Japanese fishing community.* London: Curzon.

———. 1995. *Fishing villages in Tokugawa Japan.* London: Curzon; and Honolulu: University of Hawai'i Press.

———, and Pamela Asquith. 1997. Japanese perceptions of nature: Ideals and illusions. In *Japanese images of nature: Cultural perspectives,* eds. P. Asquith and A. Kalland. Richmond, Surrey: Curzon Press.

———, and Brian Moeran. 1992. *Japanese whaling: End of an era?* London: Curzon.

Kamiyama M., Matsuzaki S., et al. 2000. Shusshō-jisei-hi no hendō to daioki-shin-osen no genjō to no chīki-teki taihi (Comparative investigation of regional sex ratio at birth and dioxin pollution). Paper presented at the Third Annual Meeting of the Japan Society of Endocrine Disruptor Research, 15 December 2000, Yokohama, Japan.

Karasawa Kōichi. 2003. *Karasu ha dore hodo kashikoi ka: Toshichō no tekiō senryaku* (How clever are *karasu?* Adaptation strategies of the city bird). Tokyo: Chūō Kōron Shinsha.

Kerr, Alex. 2001. *Dogs and demons: Tales from the dark side of Japan.* New York: Hill & Wang.

Kertzer, David I. 1988. *Ritual, politics, and power.* New Haven, CT: Yale University Press.

Kirby, Peter Wynn. 2004. Getting engaged: Pollution, toxic illness, and discursive shift in a Tokyo community. In *Confronting environments: Local understanding in a globalizing world,* ed. J.G.Carrier. Walnut Creek, CA: AltaMira.

———. 2009a. Lost in "space": An anthropological approach to movement. In *Boundless worlds: An anthropological approach to movement,* ed. Peter Wynn Kirby. Oxford: Berghahn.

———. 2009b. Toxins without borders: Interpreting spaces of contamination and suffering. In *Boundless worlds: An anthropological approach to movement,* ed. Peter Wynn Kirby. Oxford: Berghahn.

Knight, John. 1997. The soil as teacher: Natural farming in a mountain village. In *Japanese images of nature: Cultural perspectives,* eds. P.Asquith and A.Kalland. Richmond, Surrey: Curzon Press.

Kobayashi Shigeru. 1983. *Nihon shi'nyō mondai genryūkō* (A consideration of the origins of Japan's problems with night soil). Tokyo: Akashi Shoten.

Kōmeitō. 2009. *New Kōmeitō 2009 manifesto: Key policy goals and initiatives.* Summarized English version. (http://www.komei.or.jp/en/policy/manifesto09.pdf [accessed September 2009])

Kondo, Dorinne. 1990. *Crafting selves: Power, gender, and discourses of identity in a Japanese workplace.* Chicago: University of Chicago Press.

Kyburz, Josef. 1997. Magical thought at the interface of nature and culture. In *Japanese images of nature: Cultural perspectives,* eds. P.Asquith and A.Kalland. Richmond, Surrey: Curzon Press.

LaFleur, William. 1992. *Liquid life: Abortion and Buddhism in Japan.* Princeton: Princeton University Press.

La Fontaine, Jean. 1998. *Speak of the Devil: Tales of satanic abuse in contemporary Britain.* Cambridge: Cambridge University Press.

Lesbirel, S.Hayden. 1998. *NIMBY politics in Japan: Energy siting and the management of environmental conflict.* Ithaca, NY: Cornell University Press.

Lévi-Strauss, Claude. 1968. *L'Origine des manières de table* (The origin of table manners). Paris: Plon.

Lincoln, Edward J. 1988. *Japan: Facing economic maturity.* Washington, DC: The Brookings Institution.

Lock, Margaret. 1980. *East Asian medicine in urban Japan.* Berkeley: University of California Press.

Lynch, Kevin. 1960. *The image of the city.* Cambridge, MA: MIT Press.

MacCormack, Carol, and Marilyn Strathern, eds. 1980. *Nature, culture and gender.* Cambridge: Cambridge University Press.

Macfarlane, Alan. 1997. *The savage wars of peace: England, Japan and the Malthusian trap.* Oxford: Basil Blackwell.

Mackie, Vera. 1988. Division of labour: Multinational sex in Asia. In *The Japanese trajectory: Modernization and beyond,* eds. Gavan McCormack and Yoshio Sugimoto. Cambridge: Cambridge University Press.

MacNeill, Jim, Peter Winsemius, and Taizo Yakushiji. 1991. *Beyond interdepen-*

dence: The meshing of the world's economy and the earth's ecology. Oxford: Oxford University Press.

Malinowski, Bronislaw. 1922 (1961). *Argonauts of the western Pacific: An account of native enterprise and adventure in the archipelagoes of Melanesian New Guinea.* New York: Dutton.

Malkki, Liisa H. 1995. *Purity and exile: Violence, memory, and national cosmology among Hutu refugees in Tanzania.* Chicago: University of Chicago Press.

Marzluff, John M., and Tony Angell. 2005. *In the company of crows and ravens.* New Haven, CT: Yale University Press.

Matsuda Michio. 2000. *Karasu, naze osō? Toshi ni sumu yasei* (Why do *karasu* attack? The wild that inhabits the city). Tokyo: Kawade Shobō Shinsha.

———. 2006. *Karasu ha naze Tokyo ga suki na no ka?* (Why do *karasu* love Tokyo?). Tokyo: Heibonsha.

Matsuhara Iwao. 1964. *On life and nature in Japan.* Tokyo: Hokuseido Press.

Matsusaka Yoshihisa Tak. 2001. *The making of Japanese Manchuria, 1904–1932.* Cambridge, MA: Harvard University Press.

Matsushita Kazuyo. 2002. *18-nin no wakamono-tachi ga kataru: Buraku no aidentiti.* (Eighteen youths discuss buraku identity). Osaka: Buraku Kaihō Jinken Kenkyūjo.

Matsutō Toshihiko and Tanaka Nobutoshi. 1993. Toshi gomi kanri no tame no haikibutsu tōkei kairyō ni kansuru kenkyū (Study on the nationwide waste statistics in Japan for integrated solid waste management). *Haikibutsu Gakkaishi* (Journal of the Japan Society of Waste Management Experts) 4 (1):10–18.

Matsuzaki S. and M. Yamazaki (2000). Distribution and fluctuation of male ratios at birth in Japanese prefectures and municipalities during 1974–1997. Paper presented at the Third Annual Meeting of the Japan Society of Endocrine Disruptor Research, 15 December 2000, Yokohama, Japan.

Mayer, S. J. 1997. *A preliminary view and evaluation of scientific whaling from 1986 to 1996.* A report for the Whale and Dolphin Conservation Society. Bath: Whale and Dolphin Conservation Society.

McCormack, Gavan. 1996. *The emptiness of Japanese affluence.* Armonk, NY: M. E. Sharpe.

———. 2002. Breaking the Iron Triangle. *New Left Review* 13:5–23.

McKean, Margaret. 1981. *Environmental protest and citizen politics in Japan.* Berkeley: University of California Press.

Melosi, Martin V. 2001. *Effluent America: Cities, industry, energy, and the environment.* Pittsburgh, PA: University of Pittsburgh Press.

———. 2005. *Garbage in the cities: Refuse, reform, and the environment.* Rev. ed. Pittsburgh. PA: University of Pittsburgh Press.

Minami Hiroshi. 1994. *Nihonjinron: Meiji kara konnichi made* (*Nihonjinron:* From Meiji to the present). Tokyo: Iwanami Shoten.

Ministry of Agriculture, Forestry and Fisheries. 2006. *Baiomasu nippon sōgō senryaku.* (Biomass Nippon integrated strategy). Tokyo: Ministry of Agriculture, Forestry and Fisheries. (http://www.maff.go.jp/j/biomass/pdf/h18_senryaku.pdf)

Ministry of Environment. 2007a. Daiokishin-rui no haishutsuryō no mokuroku (haishutsu inbentorii). (Dioxin emissions inventory). Tokyo: Ministry of Environment.

———. 2007b. 2006 daiokishin-rui ni kakaru kankyō chōsa kekka (Results of the environmental survey regarding dioxins). Tokyo: Ministry of Environment.

Ministry of Health and Welfare. 1999. *Bonyū-chū no daiokishin-rui ni kansuru chōsa* (Dioxin breast milk survey). Tokyo: Ministry of Health and Welfare.

Ministry of Land, Infrastructure, Transport, and Tourism. 2005. *Kensetsu risaikuru ni kan suru kongo no dōkō* (Coming trends regarding construction-related waste recycling). Tokyo: Ministry of Land, Infrastructure, Transport, and Tourism.

———. 2006. *2005 kensetsu fukusanbutsu jittai chōsa kekka ni tsuite* (Results of the 2005 survey regarding the true state of construction waste byproducts). Tokyo: Ministry of Land, Infrastructure, Transport, and Tourism.

Miyake Yoshiko. 1991. Doubling expectations: Motherhood and women's factory work under state management in Japan in the 1930s and 1940s. In *Recreating Japanese women, 1600–1945,* ed. G. Bernstein. Berkeley: University of California Press.

Miyata Noboru. 1977. *Minzoku shūkyōron no kadai* (On folk religious discourse). Tokyo: Miraisha.

———. 1996. *Kegare no minzokushi: Sabetsu no bukateki yōin* (An ethnology of pollution: Discrimination's primary cultural factors). Kyoto: Jimbun Shoin.

Mocarelli, P., P. Gerthoux, et al. 2000. Paternal concentrations of dioxin and sex ratio of offspring. *Lancet* 355:1858–1863.

Moeran, Brian. 1985. *Ōkubo diary: Portrait of a Japanese valley.* Stanford: Stanford University Press.

———. 1995. Reading Japanese in Katei Gahō: The art of being an upperclass [sic] woman. In *Women, media and consumption in Japan,* eds. L. Skov and B. Moeran. Richmond, Surrey: Curzon.

———. 1996. *A Japanese advertising agency: An anthropology of media and markets.* Richmond, Surrey: Curzon; and Honolulu: University of Hawai'i Press.

———, and Lise Skov. 1997. Mount Fuji and the cherry blossoms: A view from afar. In *Japanese images of nature: Cultural perspectives,* eds. P. Asquith and A. Kalland. Richmond, Surrey: Curzon Press.

Moon Opkyo. 1997. Marketing nature in rural Japan. In *Japanese images of nature: Cultural perspectives,* eds. P. Asquith and A. Kalland. Richmond, Surrey: Curzon Press.

Morishita Jōji. 2006. Multiple analysis of the whaling issue: Understanding the dispute by a matrix. *Marine Policy* 30:802–808.

Morris-Suzuki, Tessa. 1991. Concepts of nature and technology in pre-industrial Japan. *East Asian History* 1:81–96.

Morse, Edward. 1961 (1886). *Japanese homes and their surroundings.* New York: Dover.

Mumford, Lewis. 1961. *The city in history: Its origins, its transformations, and its prospects.* New York: Harcourt, Brace.

Murakami Haruki. 2000. *Underground: The Tokyo gas attack and the Japanese psyche.* Trans. Alfred Birnbaum and Philip Gabriel. London: Harvill.

Nakane Chie. 1984 (1970). *Japanese society.* New York: Penguin.

Nakasugi Osami. 1985. Gomi shori gyōsei ni okeru jōhō kanri (Information management in waste disposal administration). *Toshi Seisō* (Journal of the Japan Waste Management Association) 38 (145):127–134.

Namihira Emiko. 1977. *Hare, ke* and *kegare:* The structure of Japanese folk belief. Ph.D. diss., Austin: The University of Texas.

———. 1984. *Kegare no kōzō* (The structure of ritual pollution). Tokyo: Seidosha.

———. 1985. *Kegare* (Ritual pollution). Tokyo: Tōkyōdō.

Nature Conservation Bureau. 2001. *Jichitai tantōsha no tame no karasu taisaku manuaru (A manual for solving crow issues for local government officials).* Tokyo: Ministry of the Environment. (http://www.env.go.jp/nature/karasu-m/index.html)

Nelson, John K. 1996. *A year in the life of a Shinto shrine.* Seattle: University of Washington Press.

New York Times. 2000, 30 March. Tokyo journal: An upstart governor takes on Japan's mandarins. By Howard W. French.

———. 2008, 7 May. Kagoshima journal: Japan fights crowds of crows. By Martin Fackler.

Nickum, James E., Aoyagi-Usui Midori, and Otsuka Takashi. 2003. Environmental consciousness in Japan. *Southeast Asian studies* 41 (1):36–58.

Ninomiya S. 1933. An inquiry concerning the origin, development, and present situation of the Eta in relation to the history of social classes in Japan. *Transactions of the Asiatic Society of Japan* 10:47–154.

Norbeck, Edward. 1965. *Changing Japan.* New York: Holt, Rinehart, and Winston.

Nordqvist, Joakim. 2006. *Evaluation of Japan's Top Runner Programme.* Utrecht: AID-EE.

Nygren, Anja. 1999. Local knowledge in the environment-development discourse. *Critique of Anthropology* 19 (3): 267–288.

O'Bryan, Scott. 2000. Growth solutions: Economic knowledge and problems of capitalism in post-war Japan, 1945–1960. Ph.D. diss., Department of History, Columbia University.

———. 2009. *The Growth Idea: Purpose and Prosperity in Postwar Japan.* Honolulu: University of Hawai'i Press.

Observer, The. 2001, 24 June. Special report: "You eat cows and pigs, so why can't we eat whales?" By Anthony Browne.

OED. 1989. *The Oxford English Dictionary.* Second ed. Oxford: Oxford University Press.

Oguma Eiji. 2002. *A genealogy of "Japanese" self-images.* Trans. David Askew. Melbourne: Trans Pacific Press.

Ōhashi Tetsurō. 2004. Production and technology of iron and steel in Japan during 2003. *ISIJ International* 44 (6):941–956.

Ohno Taiichi. 1988. *Toyota production system: Beyond large-scale production.* New York: Productivity Press.

Ohnuki-Tierney, Emiko. 1984. *Illness and culture in contemporary Japan: An anthropological view.* Cambridge: Cambridge University Press.

———. 1987. *The monkey as mirror: Symbolic transformations in Japanese history and ritual.* Princeton: Princeton University Press.

———. 1993. *Rice as self: Japanese identities through time.* Princeton: Princeton University Press.

———. 2002. *Kamikaze, cherry blossoms, and nationalisms: The militarization of aesthetics in Japanese history.* Chicago: University of Chicago Press.

Ohsako, S., Y. Miyabara, et al. 2001. Maternal exposure to a low dose of 2,3,7,8-tetrachlorodibenzo-p-dioxin (TCDD) suppressed the development of reproductive organs of male rats: dose-dependent increase of mRNA levels of 5a-reductase type 2 in contrast to a decrease of androgen receptor in the pubertal ventral prostate. *Toxicological Sciences* 60 (1):132–143.

Okuyama Masaki. 2003. Administrative measures against crows. *Global Environmental Research* 7 (2):199–205.

Ooms, Herman. 1996. *Tokugawa village practice: Class, status, power, law.* Berkeley: University of California Press.

Ortner, S. 1974. Is female to male as nature is to culture? In *Woman, culture and society,* eds. M. Rosaldo and L. Lamphere. Stanford: Stanford University Press.

———, and H. Whitehead, eds. 1981. *Sexual meanings: The cultural construction of gender and sexuality.* Cambridge: Cambridge University Press.

Parker, Robert. 1983. *Miasma: Pollution and purification in early Greek religion.* Oxford: Clarendon.

Perry, Linda. 1975. Being socially anomalous: Wives and mothers without husbands. In *Adult episodes in modern Japan,* ed. D. Plath. Leiden: E. J. Brill.

Petryna, Adriana. 2002. *Life exposed: Biological citizens after Chernobyl.* Princeton, NJ: Princeton University Press.

Pevsner, Nikolaus. 1956. *The Englishness of English art.* London: Architectural Press.

———. 1968. *The sources of modern architecture and design.* London: Thames & Hudson.

Pezeu-Massabuau, Jacques. 1981. *La maison Japonaise* (The Japanese house). Paris: Publications Orientalistes de France.

Pharr, Susan. 1990. *Losing face: Status politics in Japan*. Berkeley: University of California Press.

Problems with Nuclear Deterioration Research Society. 2008. *Maru de genpatsu nado nai ka no yō ni* (Let's imagine we got rid of nuclear power). Tokyo: Gendai Shokan.

Pyle, Kenneth. 1973. The technology of Japanese nationalism: The local improvement movement, 1900–1918. *Journal of Asian Studies* 33 (1):51–65.

Rathje, William, and Cullen Murphy. 2001. *Rubbish! The archaeology of garbage*. Tucson: University of Arizona Press.

Reader, Ian. 1991. *Religion in contemporary Japan*. Basingstoke: Macmillan.

———. 2000. *Religious violence in contemporary Japan: The case of Aum Shinrikyo*. Richmond, Surrey: Curzon.

Reid, Donald. 1991. *Paris sewers and sewermen: Realities and representations*. Cambridge, MA: Harvard University Press.

Reiter, Rayna. 1975. *Toward an anthropology of women*. New York: Monthly Review Press.

Retherford, Robert D., Naohiro Ogawa, and Rikiya Matsukura. 2001. Late marriage and less marriage in Japan. *Population and Development Review* 27:65–192.

Roberts, Glenda S. 2002. Pinning hopes on angels: Reflections from an aging Japan's urban landscape. In *Family and social policy in Japan: Anthropological approaches*, ed. R. Goodman. Cambridge: Cambridge University Press.

Robertson, Jennifer. 1991. *Native and newcomer: Making and remaking a Japanese city*. Berkeley: University of California Press.

———. 1998. It takes a village: Internationalization and nostalgia in postwar Japan. In *Mirror of modernity: Invented traditions of modern Japan*, ed. S. Vlastos. Berkeley: University of California Press.

Rohlen, Thomas P. 1974. *For harmony and strength: Japanese white-collar organization in anthropological perspective*. Berkeley: University of California Press.

Roman, B., and R. Peterson. 1998. Developmental male reproductive toxicology of 2,3,7,8-tetrachlorodibenzo-p-dioxin (TCDD) and PCBs. In *Reproductive and developmental toxicology*, ed. K. Korach. New York: Marcel Dekker.

Sakurai Tokutarō. 1970. *Nihon minkan shinkōron* (A discussion of Japanese folk beliefs). Tokyo: Kōbundō.

———, ed. 1988. *Nihon minzoku no dentō to kōzō: Shin-minzokugaku no kōsō* (Japanese folk tradition and creation: Concepts of the new folklore). Tokyo: Kōbundō.

Sansom, George. 1964. *A history of Japan, 1615–1867*. London: Cresset Press.

Sassen, Saskia. 1991. *The global city: New York, London, Tokyo*. Princeton, NJ: Princeton University Press.

Sato Jin. 2007. Formation of the resource concept in Japan: Post-war efforts in knowledge integration. *Sustainability science* 2 (2):151–158.

Scanlan, John. 2005. *On garbage.* London: Reaktion Books.

Schama, Simon. 1995. *Landscape and memory.* London: Fontana Press.

Science. 2007, 26 April. News focus: Killing whales for science? A storm is brewing over plans to expand Japan's scientific whaling program. *Science* 316:532–534. By Virginia Morell.

Seidensticker, Edward. 1991a (1983). *Low city, high city: Tokyo from Edo to the earthquake.* Cambridge, MA: Harvard University Press.

———. 1991b (1990). *Tokyo rising: The city since the great earthquake.* Cambridge, MA: Harvard University Press.

Sennett, Richard. 1994. *Flesh and stone: The body and the city in Western civilization.* London: Penguin.

Shiga Shigetaka. 1937 (1894). *Nihon fūkeiron* (Discourse on Japanese scenery). Tokyo: Iwanami Shoten.

Shiller, Robert J. 2000. *Irrational exuberance.* Princeton, NJ: Princeton University Press.

Shinjuku Ward Waste Office. 2008. *Shigen/gomi no atarashii wakekata/dashikata* (A new way of sorting and disposing of waste and recyclables). Tokyo: Shinjuku Ward Waste Office.

Sippel, P. 1998. Abandoned fields: Negotiating taxes in the Bakufu Domain. *Monumenta Nipponica* 53: 197–224.

Skov, Lise, and Brian Moeran, eds. 1995. *Women, media and consumption in Japan.* Richmond, Surrey: Curzon.

Smith, Anthony D., ed. 1992. *Ethnicity and nationalism.* Leiden: Brill.

———. 2000. *The nation in history: Historiographical debates about ethnicity and nationalism.* Cambridge: Polity Press.

Smith, Colin. 2006. After affluence: Freeters and the limits of new middle class Japan. Ph.D. diss., Department of Anthropology, Yale University

Smith, Eric Alden, and Mark Wishnie. 2000. Conservation and subsistence in small-scale societies. *Annual Review of Anthropology* 29:493–524.

Smith, Robert. 1978. *Kurusu: The price of progress in a Japanese village, 1951–1975.* Folkestone: Dawson.

Sprague, David S. 2005. This land is your land: Reintroducing Japanese landscape to contemporary Japan with the Jinsoku Sokuzu. Paper presented at the Anthropologists of Japan in Japan (AJJ) Conference, Sophia University, Tokyo, 6 November.

———, T. Goto, and H. Moriyama. 2000. GIS analysis using the Rapid Survey Map of traditional agricultural land use in the early Meiji Era. *Randosukeepu kenkyū* (Journal of the Japanese Institute of Landscape Architecture) 63:771–774. (In Japanese.)

Srinivas, Mysore Narasimhachar. 1978 (1962). *Caste in modern India, and other essays.* London: JK Publishers.

Steger, Brigitte. 2003. Getting *away* with sleep: Social and cultural aspects of dozing in Parliament. *Social Science Japan Journal* 6 (2):181–197.

Stewart, Michael. 1997. *The time of the Gypsies.* Boulder, CO: Westview Press.

Stoett, Peter J. 1997. *The international politics of whaling.* Vancouver: University of British Columbia Press.

Strasser, Susan. 1999. *Waste and want: A social history of trash.* New York: Henry Holt.

Strathern, Marilyn. 1992. *After nature: English kinship in the late twentieth century.* Cambridge: Cambridge University Press.

Süddeutsche Zeitung Magazin. 2001, 23 February. Cover.

Sugita Shōei. 2004. *Karasu: Omoshiro seitai to kashikoi fusegikata (Karasu:* Interesting ecology and clever prevention). Tokyo: Rural Culture Association.

Sumitomo K., and K. Itakura. 1998. *Sabetsu to meishin: Hisabetsu buraku no rekishi* (Discrimination and superstition: A history of buraku discrimination). Tokyo: Kasetsusha.

Süskind, Patrick. 1985. *Perfume: The story of a murderer.* London: Penguin.

Suzuki Hikaru. 2000. *The price of death: The funeral industry in contemporary Japan.* Stanford, CA: Stanford University Press.

Swan, S., E. Elkin, and L. Fenster. 1997. Have sperm densities declined: A reanalysis of global trend data. *Environmental Health Perspectives* 105:1228–1232.

Tanaka Yuki. 1988. Nuclear power and the labor movement. In *The Japanese trajectory: Modernization and beyond,* eds. Gavan McCormack and Yoshio Sugimoto. Cambridge: Cambridge University Press.

Taussig, Michael. 2003. Miasma. In *Culture and waste: The creation and destruction of value,* eds. G. Hawkins and S. Muecke. Lanham, MD: Rowman & Littlefield.

Tellenbach, H., and B. Kimura. 1989. The Japanese concept of "nature." In *Nature in Asian traditions of thought: Essays in environmental philosophy,* eds. J. B. Callicott and R. T. Ames. Albany: State University of New York Press.

Thompson, John. 1995. *The media and modernity: A social theory of the media.* Cambridge: Polity Press.

Thompson, Michael. 1979. *Rubbish theory: The creation and destruction of value.* Oxford: Oxford University Press.

Tokyo Bureau of Environment. 1999. *Nakusō daiokishi.* (Let's get rid of dioxins). Tokyo: Tokyo Bureau of Environment.

———. 2006a. *The Environment in Tokyo 2005.* Tokyo: Tokyo Bunkyūdō.

———. 2006b. *Tōkyō-to kankyō hakusho 2006* (Tokyo's environmental white paper 2006). Tokyo: Tokyo Bunkyūdō.

Tokyo Shimbun. 2005, 5 December. Karasu ha kīro ni yowai? Tokushu busshitsu haigō no gomi-bukuro de kōka. (Are *karasu* weak with yellow? The effectiveness of a garbage bag with a special combination of materials).

Totman, Conrad. 1989. *The green archipelago: Forestry in preindustrial Japan.* Berkeley: University of California Press.

Tsuda Takeyuki. 2003. *Strangers in the homeland: Japanese Brazilians return migration in transnational perspective.* New York: Columbia University Press.

Tsujimura A., T. Iwamoto, et al. 2000. Ōsaka-chiku ni okeru seijō dansei no seishokuki-nō chōsa (Sexual function of fertile Japanese men in Osaka area). Paper presented at the Third Annual Meeting of the Japan Society of Endocrine Disruptor Research, 15 December 2000, Yokohama, Japan.

Tsukuba, H. 1969. *Beishoku, nikushoku no bunmei* (Rice-eating civilization, meat-eating civilization). Tokyo: Nippon Hōsō Shuppan Kyōkai.

Turner, Victor. 1967. *The forest of symbols: Aspects of Ndembu ritual.* Ithaca, NY: Cornell University Press.

———. 1969. *The ritual process: Structure and anti-structure.* Chicago: Aldine.

Ueta M. et al. 2003. Population change of jungle crows in Tokyo. *Global Environmental Research* 7 (2):131–137.

Ugumori Taoru. 2006. *Mujirushi Ryōhin no fushigi* (The wonder of Muji). Tokyo: Pie Books.

United Nations Environment Programme. 1999. *Dioxin and furan inventories: National and regional emissions of PCDD/PCDF.* Geneva: United Nations Environment Programme.

———. 2004. *The Stockholm convention on persistent organic pollutants.* Geneva: United Nations Environment Programme. (http://www.pops.int/documents/convtext/convtext_en.pdf)

United Nations Food and Agriculture Organization. 2006. *Livestock's long shadow: Environmental issues and options.* Rome: UNFAO

United Nations General Assembly. 1983. UNGA Resolution 38/161 (19 December 1983), reprinted in *GAOR,* 37th Session, Supplement No. 47 (A/38/47):131–132.

———. 1987. UNGA Resolution 42/187 (11 December 1987), Report of the World Commission on Environment and Development. New York: United Nations.

Van Gennep, Arnold. 1965 (1908). *The rites of passage.* Trans. Monika B. Vizedom and Gabrielle L. Caffee. Chicago: University of Chicago Press.

Vogel, Ezra. 1963. *Japan's new middle class: The salary man and his family in a Tokyo suburb.* Berkeley: University of California Press.

Watanabe Kohei. 2003. The management and recycling of household waste in England and Japan: A comparative study. Ph.D. diss., University of Cambridge.

Watanabe Yonehide. 2006. *Mujirushi Ryōhin no 'kaikaku': Naze mujirushi ryōhin yomigaetta no ka?* (Muji's turnaround: How Muji resurrected itself). Tokyo: Shougyoukai.

Watanabe Zenjirou. 1991. *Kindai nihon toshi kinkō nōgyōshi* (A history of agriculture in the outskirts of cities in modern Japan). Tokyo: Ronsōsha.

Watsuji Tetsurō. 1961. *A climate: A philosophical study.* Trans. Geoffrey Bownas. Tokyo: Japanese Government Printing Bureau.

———. 1962. *Watsuji Tetsurō zenshū.*(The collected works of Watsuji Tetsurō). Vol. 8. Tokyo: Iwanami Shoten.

———. 1975 (1935). *Fūdo.* (Milieux). Tokyo: Iwanami Shoten.

Weiner, Michael, ed. 1997. *Japan's minorities: The illusion of homogeneity.* London: Routledge.

Wendelken, Cherie. 2000. Pan-Asianism and the pure Japanese thing: Japanese identity and architecture in the late 1930s. *positions: East Asia Cultures Critique* 8 (3):819–828.

White, J. W. 1988. State growth and popular protest in Tokugawa Japan. *Journal of Japanese Studies* 14:1–25.

Williams, Raymond. 1973. *The country and the city.* London: Chatto & Windus.

———. 1988 (1976). Keywords: A vocabulary of culture and society. London: Fontana Press.

World Commission on Environment and Development. 1987. *Our common future.* (Report of the World Commission on Environment and Development). Oxford: Oxford University Press.

World Health Organization. 1999. *Dioxins and their effects on human health.* World Health Organization fact sheet No. 225. Geneva: World Health Organization.

Yatsuka Hajime. 1990. An architecture floating on the sea of signs. In *The new Japanese architecture,* ed. Botond Bognar. New York: Rizzoli.

Yokoyama Hidenori. 2007. "Exploring Tokyo Bay: Present problems and future prospects of Tokyo Bay." Tokyo Waste Landfill Management Office (http://www.jsce-int.org/civil_engineering/2007/91–4-3.pdf [accessed October 2008])

Yomiuri Shimbun. 1999a, 10 February. "Hōrensō daiokishin, tsūjyō reberu-nai": JA-Tokorozawa-shi ga chōsa kekka kōhyō (Dioxin level of spinach within the normal range: Tokorozawa City Agricultural Cooperative presents results of survey). Morning edition, p. 1.

———. 1999b, 10 February. Chōsa mada hanbun na no ni: Tokorozawa dake no mondai ka? (And yet the survey was still only half finished: Is this only Tokorozawa's problem [or a broader problem?). Morning edition, p. 30.

———. 1999c, 10 February. "Sanpai Ginza" shōkyakuro 30-ka-sho (Thirty facilities in the "Industrial Waste Ginza"). Morning edition, p. 30.

———. 1999d, 10 February. Ketsuchū ni kōnōdo daiokishin (High levels of dioxin in blood). Morning edition, p. 30.

———. 1999e, 10 February. Daiokishin JA-Tokorozawa chōsa: "Konna sūji de naze" (Tokorozawa Agricultural Cooperative's dioxin survey: "Why these kinds of numbers?" Citizen group voices anxiety). Morning edition, p. 30.

———. 1999f, 10 February. "Yasai to hyōgen machigai datta": Terebi-Asahi mitomeru (TV-Asahi admits "[Using] the expression 'vegetable' was a mistake"). Morning edition, p. 30.

———. 1999g, 10 February. Kagaku-busshitsu no haishutsu ya baibai: "Todokede jyōhō" kuni ga kanri (The central government will now manage the emission, buying, and selling of dangerous chemical substances). Evening edition, p. 1.

———. 1999h, 10 February. Tokorozawa-san no yasai "anzen sengen" (Vegetables from Tokorozawa "pronounced safe"). Evening edition, p. 18.

———. 1999, 27 March. Nose no shōkyaku-jō sagyō-in: Kō-nōdo ketsu-chū daiokishin (Nose incinerator workers: High dioxin level in blood). Morning edition, p. 1.

———. 1999, 30 March. Nose ya Tokorozawa jūmin, Kankyō-chō chōsa: Daiokishin ketsu-chū nōdo (Nose and Tokorozawa residents, Environment Agency study: Dioxin blood level). Morning edition, p. 38.

———. 1999, 3 August. Bonyū ni daiokishin (Dioxins in mother's milk). Morning edition, p. 1.

———. 2006, 10 March. Karasu ha naze kīro ga nigate? (Why is yellow a weakness for *karasu?*). By Okemura Kumiko.

Yoshino K. 1992. *Cultural nationalism in Japan: A sociological inquiry.* London: Routledge.

Young, Louise. 1998. *Japan's total empire: Manchuria and the culture of wartime imperialism.* Berkeley: University of California Press.

Yuasa Yasuo. 1987. The encounter of modern Japanese philosophy with Heidegger. In *Heidegger and Asian thought,* ed. G. Parkes. Honolulu: University of Hawai'i Press.

Index

abortion, 137, 157–159; and *mizuko kuyō*, 158

affluence, 15, 163

agricultural land, toxic pollution of, 52–53

air emissions. *See* air pollution.

air pollution, 5, 6; automobile exhaust, 7, 63; and laundry, 124. *See also* incineration: impromptu burning

America, 15, 196, 219n.17; environmental reputation of, 9; unsustainable practices in, 18–19

animals: carcasses of, 18, 46, 203n.16; contaminated fish, 38; dumping of pets, 7; and environmental hormones, 66–67; "ethical" treatment of whales, 171–172; relations with humans, 13; and vermin, 86

anthropology, 193–194, 198–199; and dangerous fieldwork, 60–61

Askew, David, 209n.13, 213n.18

asthma, 66

atomic bombings, discrimination against survivors of, 20, 140

atopī (atopic dermatitis), 66

Azuma Disease: dangerous chemicals and, 185; hearings on, 46, 47, 184, 205n.1; ill-defined nature of threat of, 126–127; knowledge transmission via rumor, 61; misconceptions about outside of Izawa, 61, 64; naming strategy of, 51, 205n.1; self-revulsion of sufferers, 123; social frictions caused by, 130; solidarity created by, 122–123; symptoms of, 28–30, 123; Tokyo's partial acknowledgment of, 184–185; unknown dangers of, 123

banks, 11, 208n.5

bathhouses, 120, 215n.18

Baudrillard, Jean, 15, 17, 202n.9, 203n.18

beauty, Japanese ideas of, 23, 83–84, 174, 208n.4

Bestor, T., 5

Bhopal, 18, 38, 50, 194

biomass, 190–191, 194. *See also* waste: as energy

birth defects, 11, 64–65, 145–147, 215n.10

blame, 42, 146

bodies: as antennae, 40; "innate" dispositions of, 41; toxic damage of, 46

breast milk, contamination of, 11, 64–65, 145–147, 151

Brundtland Commission, 162, 164

Bubble years, 31, 134, 153, 163, 164–165, 168, 183; and waste, 87, 88

bulky waste *(sodai gomi),* 88; as slang term, 89–90

burakumin (customary "untouchables"), 103, 211n.2, 211n.3, 212n.8; "*burakumin*" community in Horiuchi, 118–120; discrimination against,

About the Author

Peter Wynn Kirby is Senior Lecturer in the Anthropology of Japan at Oxford Brookes University, Oxford, UK, and a research fellow at the Ecole des Hautes Etudes en Sciences Sociales (EHESS), Paris. He holds a Ph.D. in anthropology from the University of Cambridge. Kirby's edited book *Boundless Worlds: An Anthropological Approach to Movement* was published in 2009.